Venture Capital and Corporate Venture Capital in Financing Biotech Companies

European University Studies

Europäische Hochschulschriften
Publications Universitaires Européennes

Series V
Economics and Management

Reihe V   Série V
Volks- und Betriebswirtschaft
Sciences économiques, gestion d'entreprise

Vol./Bd. 3404

PETER LANG
Frankfurt am Main · Berlin · Bern · Bruxelles · New York · Oxford · Wien

Stefanie B. Hochhold

Venture Capital and Corporate Venture Capital in Financing Biotech Companies

PETER LANG
Internationaler Verlag der Wissenschaften

Bibliographic Information published by the Deutsche Nationalbibliothek
The Deutsche Nationalbibliothek lists this publication in the Deutsche Nationalbibliografie; detailed bibliographic data is available in the internet at http://dnb.d-nb.de.

Zugl.: München, Univ., Diss., 2011

D 19
ISSN 0531-7339
ISBN 978-3-631-61874-5
© Peter Lang GmbH
Internationaler Verlag der Wissenschaften
Frankfurt am Main 2012
All rights reserved.

All parts of this publication are protected by copyright. Any utilisation outside the strict limits of the copyright law, without the permission of the publisher, is forbidden and liable to prosecution. This applies in particular to reproductions, translations, microfilming, and storage and processing in electronic retrieval systems.

www.peterlang.de

Acknowledgements

Even though the main work on this dissertation was done reclusively it would have not been possible without the support, effort and guidance of many other people. I want to express my sincere appreciation and gratitude to all of them, some of them are explicitly mentioned below.

First, and foremost, I want to thank Prof. Dr. Bernd Rudolph; my doctoral advisor; for the freedom in doing my research and his valuable advice and guidance every time it was requested. I also want to thank all my former colleagues at the Institute of Capital Market Research and Finance for creating an animated working environment during the last four years. All of them contributed significantly to the fact that I have enjoyed almost every day of my work.

I am also indebted to Prof. Dr. Manfred Schwaiger for being my mentor during my MBR studies and becoming my thesis referee.

I would like to express special thanks to Prof. Dr. Carolin Häussler and Prof. Dr. Matthew Higgins for their helpful comments and the constructive cooperation on our joint research projects.

I also would like to thank Thomas Broichhausen, Raphaela Dreßler, Isabella Semmelmann and Melanie Wallner.

Finally, I owe gratitude to my family. To my sister Bettina Hochhold and especially to my parents, Else and Siegfried Hochhold, who always supported me unconditionally, both mentally and financially, on my academic path.

This dissertation is dedicated to my beloved grandmother Martha Brunsch († 2006) who is my great ideal in mastering life.

Table of Contents

List of Tables 11

1 **Introduction** 13
   Bibliography for Chapter 1 20

2 **Monitoring and Support in Venture Capital Financing** 23
   2.1 Introduction 23
   2.2 Level of Monitoring and Support 28
       2.2.1 Mechanisms Due to Institutional Framework 28
       2.2.2 Mechanisms Due to Venture Capitalist's Experience 32
   2.3 Methodology and Results 36
       2.3.1 Data Set 36
           2.3.1.1 Data Description 36
           2.3.1.2 Definition of Variables 37
           2.3.1.3 Control Variables 40
           2.3.1.4 Summary Statistics 42
       2.3.2 Methodology and Empirical Results 49
           2.3.2.1 Ordinal Regression Model 49
           2.3.2.2 Influence of Institutional Framework and Venture Capitalist's Experience on Monitoring 51
           2.3.2.3 Influence of Institutional Framework and Venture Capitalist's Experience on Support 52
       2.3.3 Discussion of Empirical Findings 59
   2.4 Conclusion 60
   Bibliography for Chapter 2 62

3 The Role of Venture Capital and Corporate Venture Capital in Financing Biotech Companies 71

3.1 Introduction 71

3.2 Theoretical Framework 74

    3.2.1 Venture Capital versus Corporate Venture Capital 74

    3.2.2 Marginal Utility of Support 76

3.3 Supporting Entrepreneurial Companies 79

    3.3.1 Corporate Venture Capitalist's and Venture Capitalist's Support 79

    3.3.2 Information Asymmetry 85

3.4 Methodology and Results 88

    3.4.1 Field of Study 88

    3.4.2 Data Set 90

        3.4.2.1 Data Collection and Description 90

        3.4.2.2 Definition of Variables 91

        3.4.2.3 Control Variables 92

        3.4.2.4 Descriptive Statistics 94

    3.4.3 Methodology and Empirical Results 99

        3.4.3.1 Seemingly Unrelated Bivariate Probit Model 99

        3.4.3.2 Empirical Results 100

    3.4.4 Discussion of Empirical Results 106

3.5 Conclusion 107

Bibliography for Chapter 3 109

4 Corporate Venture Capital Performance – Is There a First-Mover Advantage? ... 115

4.1 Introduction ... 115

4.2 Corporate Venture Capital Returns ... 119
4.2.1 Determinants of Corporate Venture Capital Returns ... 119
4.2.2 Timing of Corporate Venture Capital Returns ... 122
4.2.2.1 Implications of Corporate Finance Theory ... 122
4.2.2.2 Implications of Value-adding ... 123

4.3 Data Set ... 125
4.3.1 Sample Selection ... 125
4.3.2 Variables and Descriptives ... 126
4.3.2.1 Dependent Variables ... 128
4.3.2.2 Independent Variables ... 132
4.3.2.3 Control Variables ... 132

4.4 Empirical Results ... 137
4.4.1 Descriptive Statistics ... 137
4.4.2 Multivariate Regressions ... 139
4.4.2.1 Return Index ... 139
4.4.2.2 Technological Return ... 142
4.4.2.3 Financial Return ... 145

4.5 Conclusion ... 148

Bibliography for Chapter 4 ... 151

List of Tables

Table 2.1: Overview of Independent Variables and Expected Impact on Monitoring and Support 44

Table 2.2: Summary Statistics 45

Table 2.3: Pairwaise Correlation 48

Table 2.4: Impact of Institutional Framework and VC Experience on Formal Reporting 55

Table 2.5: Impact of Institutional Framework and Experience on Personal Communication 56

Table 2.6: Impact of Institutional Framework and VC Experience on Availability 57

Table 2.7: Impact of Institutional Framework and VC Experience on Advice 58

Table 3.1: Overview of Independent Variables and Expected Impact on Being VC or CVC financed. 95

Table 3.2: Descriptive Statistics 97

Table 3.3: Correlations of Independent Variables 98

Table 3.4: Seemingly Unrelated Bivariate Probit Models 102

Table 4.1: Overview of Variables 127

Table 4.2: Summary Statistics 130

Table 4.3: Four Field Matrix of Technological and Financial Return.131

Table 4.4: Correlations..136

Table 4.5: Descriptive Statistics of CVC Timing and Returns............138

Table 4.6: Return Index 141

Table 4.7: Technological Return 144

Table 4.8: Financial Return 147

1 Introduction

Innovative entrepreneurial firms with novel technologies often lack funding for the development and introduction of new products. Investing time and capital in these entrepreneurial companies is a hazardous decision since high informational asymmetries, which are common in a R&D intensive environment, are related to those companies. The traditional banking sector often does not want to take the related risks and due to the insufficient information public equity is also rarely available. Thus, young and small research intensive companies frequently rely on specialised financial intermediaries who focus on investing in highly innovative and independent entrepreneurial companies.

Venture capital (VC) and corporate venture capital (CVC) meet these particular issues of entrepreneurial companies, playing a dual role in financing and advising: Besides providing equity or equity linked finance to entrepreneurial companies, VC and CVC also provide value-adding activities to their portfolio companies.[1] Potential conflicts of interest between the intermediary and their portfolio company as a result of high information asymmetries are mitigated by the VC's and CVC's monitoring and support function.[2]

In order to understand which factors lead to the success of VC and CVC investments, variables which have direct as well as indirect effects on investment performance need to be analysed. With respect to the VC industry some researchers argue direct effects, such as differences in monitoring and support, levels of syndication and the experience of VC investors are important criteria for VC

---

1 For VC see Gorman/ Salman (1989), for CVC see Dushnitsky/ Lenox (2006).
2 See Tykvova (2003).

fund performance.[3] Other researchers focus on indirect effects: They suggest that the institutional framework and environmental factors such as market rigidities, efficiency of initial public offerings (IPO) markets, or fiscal environments, explain a significant share of variation within VC performance.[4]

Even though CVC finances entrepreneurial companies alongside independent VCs, CVC funds differ from VC funds in their organizational and incentive structure. According to theory the structure of VC funds, in particular the reliance on limited partnerships of finite life with substantial profit sharing is critical to success. CVC is structured as a corporate subsidiary and has a much lower incentive-based compensation. The therefore less efficient CVC fund structure may be offset by benefits emanating from superior information and complementary activities with a portfolio company.[5] When investing in young and research intensive portfolio companies, CVC may be able to select the better firms and may add greater value to their portfolio companies through using information and knowledge from their related lines of business.

For CVC investments the factors determining success in monitoring and support of entrepreneurial companies are still largely unknown. The literature concerning determinants effecting CVC performance is limited. Furthermore, CVC investments have prospects for two major performance dimensions: Strategic and financial.[6] Strategic investments seek to exploit synergies with the port-

---

3   See for example Gompers/ Lerner (2006); Hege et al. (2003); Hsu (2004).
4   See for example Jeng/ Wells (2000); Marti/ Balboa (2001); Armour/ Cumming (2004).
5   See Gompers/ Lerner (2006).
6   See for example Corporate Strategy Board (2000); MacMillan et al. (2008).

folio company, while financially motivated investments seek attractive financial returns.[7]

To shed further light on the role of VC and CVC in financing entrepreneurial companies and determinants affecting investment performance, I discuss three selected research questions in this dissertation: Determinants affecting the intensity of VC's monitoring and support activity; VC's versus CVC's value-adding contribution to their portfolio companies; and the time-dependent performance of CVC investments.

I use three independent and specific datasets to discuss these research topics. The data focuses on companies within the biotechnology industry, since this industry provides an attractive setting for studying VC and CVC investments to the same degree, and is therefore most suitable for addressing my research questions. The R&D process in the biotechnology industry is highly uncertain and complex and companies need to access a broad range of technological and human resources as well as capital. Therefore the monitoring and support activities of capital providers are important to mitigate principal agent problems resulting from high information asymmetries. Additionally, value-adding activities of VC and CVC investors are especially important for an efficient resource allocation in the biotechnology industry. A common drawback resulting from focusing on a specific industry is that my results cannot be generalized for an overall industry attempt without further consideration.

The first part, chapter 2, analyses the level of VC's monitoring and support activity. In contrast to previous studies, which focus on VC contracts and informal mechanisms by characteristics of the

---

7   See Chesbrough (2002).

entrepreneur and the portfolio company,[8] I examine particular VC financing mechanisms by differences in the national institutional framework and the experience of the VC investor. Amongst existing literature dealing with factors that determine the level of VC's value-adding activity,[9] my analysis of the influence of the institutional framework and VC's experience based on the same data set is a new approach which allows for a direct comparison of the two factors. I assume that all support activities providing value to a portfolio company increase with a market-orientation of the portfolio company's institutional environment and with an increase in VC experience. In contrast, monitoring activities that solely reduce agency costs increase with a bank-orientated financial system and decrease with an increase in VC experience.

Since the level of monitoring and support is not contractually specified and therefore not easily available, the analysed data was collected in a survey amongst 79 German and UK biotechnology companies.

I find VC experience has a positive impact on the level of support activity and the level of support increases in a market-based financial system. Furthermore, a bank-based financial system has a positive impact on the frequency of formal monitoring, whereas VC's experience is positively related to the frequency of monitoring by personal communication.

The question whether the actual value-adding ability of VC investments in contrast to CVC investments differs from a portfolio company's perspective is analysed in chapter 3. When engaging with an entrepreneurial company VC and CVC do not only com-

---

8   See for example Kaplan/ Strömberg (2003, 2004).
9   See for example Gorman/ Sahlman (1989); Manigart et al. (2002); Kaplan/ Strömberg (2003).

pete on the price of equity, but also on the level of support they can credibly provide. Previous studies claim VC distinguishes itself from CVC through their value-adding contributions to the commercial success of their portfolio company.[10] For an overall industry framework, it has been shown that when both VC and CVC are invested in a portfolio company at the same time, the value-adding contribution of VC is focused on managerial related advice, whereas CVC focuses on providing research related resources and advice. These value-adding contributions have been tested ex post, after the investment of VC and CVC has already taken place.[11]

Consequently, if the capital allocation from a portfolio company's perspective is efficient the likelihood of receiving VC and / or CVC should be dependent on an entrepreneurial company's specific need for value-adding activities.

I utilise survey data from 174 European biotechnology companies containing VC as well as CVC investments. In a simultaneous empirical framework, I analyse whether VC and CVC act as supplements or as complements in their role of meeting the portfolio companies' needs in adding value. I find VC is more likely if the company has a general demand for managerial related advice, whereas the likelihood for CVC increases with a demand for research related resources. The results provide new empirical support for the assumption that entrepreneurial companies choose their financier ex ante, VC versus CVC, based on their company specific demand for the respective value-adding activities.

The time-dependent performance of CVC investments is analysed in chapter 4. Empirical studies seem to suggest CVC investments generally have the ability to create financial as well as tech-

---

10 See Gompers/ Lerner (2006).
11 See Maula et al. (2005) and Hellmann (2002).

nological value for the investing company.[12] Additionally, there is evidence to suggest returns to CVC investors may be superior to VC investors.[13] Research has demonstrated that early stage VC investments yield, on average, lower returns than later stage investments.[14] This is in contrast to corporate finance theory which argues that risk-averse investors should receive a premium for early stage investments due to higher investment risk. Moreover, in most industries, CVC investments are occurring in a multi-partner investment setting. That is, several CVC investors may be investing sequentially or simultaneously, at potentially different stages, in a particular portfolio company.

Given this setting, the question I address is whether CVC returns are more in line with finance theory or the existing empirical VC literature, and whether the timing of "entry" of CVC investment has an impact on a CVC investor's performance.

I argue that three sources of early- and first-mover advantages are apparent. Firstly, in line with corporate finance theory, the first and early movers should receive a premium as compensation for increasing risk due to be an early investor. Secondly, the first investor may be able to receive a "discover" premium for the capability in identifying a promising venture and sending signals to other investors, and thirdly, being the first enables to set the portfolio company on a technological path that is aligned with the main business of the CVC company, and thus, attractive in terms of technology sourcing.

Based on a dataset consisting of 689 CVC financing rounds in 300 biotechnology companies I analyse the impact of CVC's entry

---

12 See for example Allen/ Hevert (2007); Dushnitsky/ Lenox (2006).
13 See Gompers/ Lerner (2006).
14 See Cumming (2008) and Cumming/ Walz (2004).

within a multi-partner investment setting on overall CVC investment returns, technological returns and financial returns. I find a relation between the timing of a CVC investment and investor returns. Notably, my assumption of a first-mover advantage is empirically supported.

Bibliography for Chapter 1

Allen, Stephen / Hevert, Kathleen (2007): Venture Capital Investing by Information Technology Companies: Did It Pay? in: Journal of Business Venturing 22, 262 - 282.

Armour, John / Cumming, Douglas (2004): The Legal Road to Replicate Silicon Valley, Working Paper University of Cambridge No. 281, March 2004.

Corporate Strategy Board (2000): Corporate Venture Capital: Managing Equity Investments for Strategic Returns, Unpublished Working Paper, Corporate Strategy Board.

Cumming, Douglas (2008): Contracts and Exits in Venture Capital Finance, in: The Review of Financial Studies 21, 1947-1982.

Cumming, Douglas / Walz, Uwe (2004): Private Equity Returns and Disclosure Around the World, LSE Working Paper, April 2004.

Dushnitsky, Gary / Lenox, Michael (2006): When Does Corporate Venture Capital Investment Create Firm Value?, in: Journal of Business Venturing 21, 723-772.

Gompers, Paul / Lerner, Josh (2006): The Venture Capital Cycle, 2nd Edition, MIT Press.

Gorman, Michael / Sahlman, William (1989): What Do Venture Capitalists Do?, in: Journal of Business Venturing 4, 231-248.

Hege, Ulrich / Palomino, Frédéric / Schwienbacher, Armin (2003): Determinants of Venture Capital Performance: Europe and the United States, Working Paper University of Amsterdam, November 2003.

Hellmann, Thomas (2002): A Theory of Strategic Venture Investing, in: Journal of Financial Economic 64, 285-314.

Hsu, David (2004): How Much Do Entrepreneurs Pay for Venture Capital Affiliation, in: Journal of Finance 59, 1805-1844.

Jeng, Leslie / Wells, Philippe (2000): The Determinants of Venture Capital Funding: An Empirical Analysis, in: Journal of Corporate Finance, 241-289.

Kaplan, Steven / Strömberg, Per (2003): Financial Contracting Theory Meets the Real World: An Empirical Analysis of Venture Capital Contracts, in: Review of Economic Studies 70, 281-315.

Kaplan, Steven / Strömberg, Per (2004): Characteristics, Contracts, and Actions: Evidence from Venture Capital Analysis, in: The Journal of Finance 59, 2177-2210.

Manigart, Sophie / Baeyens, Kathleen / Van Hyfte, Wim (2002): The Survival of Venture Capital Backed Companies, in: Venture Capital 4, 103 -124.

Marti, José / Balboa, Marina (2001): Determinants of Private Equity Fund Raising in Western Europe, SSRN Working Paper, 2001.

Maula, Markku / Autio, Erkko / Murray, Gordon (2005): Corporate Venture Capitalists and Independent Venture Capitalists: What do They Know, Who do They Know and Should Entrepreneurs Care?, in: Venture Capital 7, 3-21.

MacMillan, Ian / Roberts, Edward / Livada, Val / Wang, Andrew (2008): Corporate Venture Capital (CVC): Seeking Innovation and Strategic Growth, National Institute of Standards and Technology, U.S. Department of Commerce, June 2008.

Tykvová, Tereza (2007): What Do Economists Tell Us About Venture Capital Contracts?, in: Journal of Economic Surveys 21, 65-89.

2 Monitoring and Support in Venture Capital Financing

2.1 Introduction

A Venture Capitalist (VC) is a financial intermediary primarily investing institutional capital in privately owned, early stage technology related companies, with large growth potential. These young companies can be characterized by high levels of uncertainty due to their mainly intangible assets and lack of track record.[15] As the banking sector does not want to take these extreme risks, and due to insufficient information, public equity is also rarely available, hence private equity is the most appropriate financing source. To address these special needs of portfolio companies, VCs are active investors that not only bring private equity or private equity-related capital, but also potentially relevant knowledge, business contacts, networks, reputation and strategic advice to their investments.

The relationship between a VC and its portfolio companies is characterized by high information asymmetries leading to adverse selection and moral hazard as well as high uncertainty and risk. To reduce the exposure in financing small companies in early stages, VC contracts feature special contractual covenants. VC contracts include the use of special financing instruments such as convertible securities, syndication of the investment and financing the companies in stages.

However, contracts are incomplete, as not all eventualities can be anticipated at the time of writing a contract.[16] To mitigate fur-

---

15 See Stinchcombe (1965).
16 See Hart / Moore (1999).

ther agency conflicts evolving, informal financing mechanisms[17] are needed to compensate the incompleteness of these contracts. Informal financing mechanisms, in particular monitoring and support, play an important role in financing young start-up companies and constraining potential agency costs.[18]

The specific nature of VC contracts is chosen to reduce post-contractual information asymmetries for the VC as well as for the management of the entrepreneurial company. Financial contracting plays an important role in minimizing agency costs through aligning incentives and mitigating agency conflicts between investors and entrepreneurs. There is a large body of theoretical literature modelling the VC-entrepreneur relationship as a double moral hazard problem. This problem occurs when two parties have to provide an effort that is crucial for the success of a venture, but is not ex ante contactable and barely observable. To mitigate the double moral hazard problems and limiting information asymmetries optimal contract designs, giving incentives to provide costly effort, have been analysed.[19]

Ambiguous empirical evidence suggests the level of VC support and monitoring is not solely driven by reducing information asymmetry and opportunistic behaviour, as assumed by the majority of theoretical work.[20] Another reason for the application of informal mechanisms, despite reducing agency problems within incomplete contract theory, is to add value to the portfolio company,

---

17   In this study all non contractible and qualitative mechanisms often not objectively valuable are referred to as informal financing mechanism.
18   See Gompers (1995); Gormann/ Sahlman (1989).
19   See for example Kaplan/ Strömberg (2003); Repullo/ Suarez (2004); Inderst/ Müller (2004); Schmidt (2003).
20   See Gormann/ Sahlman (1989); Sahlman (1990); Gompers/ Lerner (1999); Bottazi et al. (2008).

as many entrepreneurs usually have neither enough business experience nor the necessary networks. VC research on informal financing mechanisms includes empirical analysis of VC added value as well as comparisons of added value effects on different types of portfolio firms. Identified value-adding factors include acting as a sounding board, assistance in obtaining additional financing, recruitment of management and board members, monitoring of financial and operating performance and providing access to networks and contacts.[21]

Previous research suggests that the level of VC involvement in implementing these informal mechanisms can be influenced by three main factors. The first factor is related to portfolio company characteristics. VC involvement is dependent on the level of uncertainty and information asymmetries. The level of involvement is a trade off between mitigation agency costs and the costs of providing effort.[22] The discrepancy between suggestions in theoretical literature regarding the actual VC involvement can be explained by two other approaches. The approaches are national institutional settings in the respective country[23] and the experience of the respective VC firm.[24] The institutional framework includes a country's institutional structure, stock market performance and activity. VC experience includes the expertise of the investor and its ability to provide value-adding assistance to portfolio companies.

The institutional framework approach consists of institutional and environmental factors that generally have more indirect effects on monitoring and support activity. The institutional framework is, however, of high importance in order to create and keep a VC in-

---

21 See for example Gorman/ Sahlman (1989); Gompers/ Lerner (1999).
22 See Gompers (1995).
23 See Jeng/ Wells (2000); Marti/ Balboa (2001); Armour/ Cumming (2004).
24 See Gompers/ Lerner (2001); Hege et al. (2003); Hsu (2004).

dustry alive.[25] The VC experience approach includes factors related directly to the VC and its portfolio companies.

In contrast to the large body of literature that studies VC contracts and informal mechanisms by characteristics of the entrepreneur and the portfolio company,[26] I will examine particular VC financing mechanisms by differences in the institutional framework in Germany and the UK, as well as the experience of VC. Analysing the influence of the institutional framework, as well as VC experience, within the same data is a new approach and allows for comparison of the impact of the institutional framework and VC experience on monitoring and support activity.

The use of a data set of German and UK portfolio companies is a matter of particular interest as Germany is a leading example of a bank-based financial system, in contrast to the market-based Anglo-Saxon system.[27] Germany's financial system is commonly characterized as being bank-dominated, whereas the British financial system is considered to be market orientated. British banks play a relatively insignificant role as financial intermediaries and as providers of longer term financing to the business sector. The stock market in Britain is of great importance. The primary, as well as secondary market, are well developed in terms of its institutional and economic role which leads to a higher rate of IPOs.[28] Also the structure of the general partners differs in both countries. In Germany, banks play an important role in financing new business, as well as insurance companies and the public sector. The main sup-

---

25 See Armour/ Cumming (2004).
26 See for example Kaplan/ Strömberg (2003, 2004).
27 See Hackethal (2004).
28 See Schmidt/ Tyrell (2004).

pliers for VC in the UK are pension funds and insurance companies.[29]

Since the level of monitoring and support is not contractually specified and therefore not easily available, the analysed data was collected in a survey amongst German and British biotechnology companies. The data provides a unique international insight into the monitoring and support activity of VC financing. In this respect monitoring is defined as an instrument to reduce agency costs resulting from post-contractual uncertainty about the entrepreneurs' quality and actions. Monitoring is measured by the frequency of formal reporting and personal communication. Support is defined as the effort to enhance the chance of a successful outcome of the investment in adding value to the portfolio company and reducing agency costs to generate a monetary benefit. Support is measured by the time it takes a necessary decision by VC to be made in urgent cases, representing VC's availability, as well as the level of advice of VC in business and operational decision making. The level of monitoring and support is controlled by company determinants that might influence formal and informal communication frequency, availability and advice activity of the VC.

I find VC experience has a positive impact on the level of support activity, whereas the institutional framework influences the advice activity since the level of advice increases in a market-based financial system. Furthermore, a bank-based financial system has a positive impact on the frequency of formal reporting, whereas the VC's experience is positively related to the frequency of personal communication.

The organization of the chapters is as follows. In section 2, the contribution of the institutional framework and the experience of

---

29 See Franzke et al. (2004).

the VC industry, determining the level of monitoring and support in VC financing, are theoretically examined. The hypotheses are derived from an assumption that marginal costs of monitoring and support exist, which should not be greater than the marginal benefits in equilibrium. In section 3, the data sample is described and summarized by descriptive statistics. The impact of the institutional framework and VC experience on monitoring and support activity are tested in an ordinal regression model. Section 4 summarizes the results.

2.2 Level of Monitoring and Support

2.2.1 Mechanisms Due to Institutional Framework

The institutional framework refers to the legal and fiscal environment in which the portfolio company is located. It can be differentiated in bank and stock market orientated systems, whereby Germany is an example of a bank orientated system; the UK a stock market orientated system.[30]

Within this institutional framework a hands-on and a hands-off investment style in VC involvement can be distinguished. The hands-on approach includes extensive monitoring and support from VC; hands-off implies less monitoring and support activity. VCs in a more stock market orientated framework are traditionally more hands-on in their support, which also indicates more monitoring compared to a more hands-off approach in bank orientated systems.[31]

---

30   See Schmidt/ Tyrell (2004).
31   See Sapienza et al. (1996).

Schefzyk[32] and Gaida[33] highlight the differing characteristics of venture capital finance in Germany. VCs in Germany have developed as bank-affiliated institutions that have recruited their portfolio companies by using an internal bank distribution network. Banks and the German government operated as limited partners. This was one of the reasons for a non-competitive environment for investing in innovative ideas and entrepreneurial companies. As VC was closely related to banks, foreign VCs could not easily enter the German market. In addition, the market for foreign VC investors was not that attractive since IPO as an exit alternative was relatively rare compared to in the UK. Black and Gilson[34] assume that the German market's structure of start-up financing and the comparably low experience level of its market participants are due to the lack of a vibrant stock market. Disparities in the financing structure between German and UK companies are also due to differences in the organisation of VC. For example, VCs in the UK implement convertible securities to mitigate possible agency conflicts, whereas German VCs are often structured as silent partnerships, which are also suitable to mitigate agency conflicts in the case VC does not intend to get actively involved in its portfolio company.[35]

The legal and fiscal environment is principally defined by the government. The government can intervene directly in the VC process by funding and managing government sponsored VC funds and providing incentives or obstacles to VC investments through regulated private companies, such as banks, pension funds or in-

---

32 See Schefzyk (2000).
33 See Gaida (2002).
34 See Black/ Gilson (1998).
35 See Haagen (2008).

surance companies.[36] The government can indirectly intervene by regulating the fiscal environment. Public market access in a developed exit market is necessary for a VC industry, since exits gain higher revenues in stock market-centred financial systems.[37] The IPO market for high-risk companies in Germany is comparatively weaker than the UK market.[38]

The academic view regarding influence on stock markets is unanimous – VCs are more successful in countries with liquid stock markets. Jeng and Wells (2000) even find it surprising that VCs in countries with underdeveloped IPO markets do not make use of the more developed IPO markets in countries like the U.S. Also, the question is being raised of whether the financial contracts observed in the U.S., which are in line with VC contracting theory, are de facto optimal in other institutional environments.

Within an institutional setting the exit environment can be indirectly controlled by the government through altering the fiscal environment. In a capital market orientated environment, IPOs are more likely to be used as an exit strategy than trade sales or secondary sales. There seems to be a general belief, in academic literature, that an IPO is the most profitable and prestigious exit route for venture capitalists. Therefore, in a capital market orientated institutional framework the expected revenue achieved in exiting the portfolio company is higher.[39] Additional costs for value-adding ac-

---

36 See Leleux/ Surlemont (2003).
37 See Becker/ Hellmann (2003); Black/ Gilson (1998).
38 See Cumming (2006).
39 Gompers (1995) states that according to Venture Economics in 1988 VC-backed companies that went public yield a return for VC investors on average of 59.5 percent per year, whereas in acquisitions an average return of 15.4 percent was realised per annum.

tivity in a market based institutional framework are more likely to be compensated for by realising higher expected revenues.

This has been shown in a simple model analogous to Cumming and MacIntosh (2001), which had a similar consideration for the duration of VC investments, as well as Cumming (2006) who uses a cost-benefit analysis depending on the funds under management. VC wants to maximize its expected revenue through adding value. The projected value added or monetary benefit is a function of the monitoring and support costs, as well as reduced agency costs. At a point of optimal monitoring and support activity the projected marginal value added is equal to the projected marginal costs.

With a liquid stock market, exits of high quality companies in a market orientated environment are more profitable as IPOs are more likely. It is therefore suggested the marginal benefit increases with a market orientation of the institutional framework, whereas the marginal costs are stable. Therefore the level of value-adding activities, as well as activities to reduce agency costs, should be higher with a market orientated institutional environment.

When a VC is available at short notice, in a situation where urgent decisions are needed, possible opportunistic behaviour of the entrepreneur might be reduced and value is added with the superior management quality of the VC. Being highly available generates costs for the VC. Since the gained marginal benefit of being highly available is dependent on the market orientation of the respective institutional environment, VC availability should be higher in a stock-market orientated financial system.

Hypothesis 1 *VC availability will increase in a market-based economy.*

When giving advice to the portfolio company, the VC adds value through its superior ability in business and operational decision making. As providing advice takes resources from the VC, it is costly and the argument above applies.

> *Hypothesis 2*   *VC advice activity will increase in a market-based economy.*

The costs of personal communication in comparison to the costs of formal reporting by agreed standards are higher. On the other hand, monitoring within a personal setting not only reduces agency costs but can also add value through personal interaction. With higher marginal benefits, in a market orientated framework, monitoring becomes less formal.

> *Hypothesis 3*   *The frequency of personal communication between the portfolio company and the VC will increase in a market-based economy.*

> *Hypothesis 4*   *The frequency of formal reporting will decrease in a market-based economy.*

2.2.2 Mechanisms Due to Venture Capitalist's Experience

An often discussed success factor of VC investment is the skill set of VC firm's investment managers. There is growing literature on classifying VC differences in quality, behaviour and ability to add value to their portfolio companies.[40] The role of a VC investor spans from a more supportive function adding value to a more monitoring function reducing agency costs.

The maturity of a VC market, that is the experience of a VC, is an indicator for the likelihood of success of the portfolio company. More experienced VCs can influence companies in more efficient ways to reduce agency problems. Due to their superior ability in monitoring and support they are more likely to contain agency problems through non-contractual channels.[41] Experienced VCs also benefit from greater network sharing resources and know-how. Consequently, adding value for more experienced VCs is less costly. Sorensen (2008) shows that, even when controlling for selection effects, portfolio companies backed by experienced VCs are more likely to go public.

Campbell and Frye (2008) argue a possible reason for a more active involvement of experienced VCs would be to protect their valuable reputation as successful investors. Their argument is contrary to the "grandstanding-hypothesis"[42] of Gompers (1995, 1996) who find evidence that younger VCs get more involved than older ones to build up reputation as a successful VC.

With increasing experience, VCs become more efficient in monitoring and supporting their portfolio companies. This leads to a decrease in the marginal costs of VC's involvement. Furthermore,

---

40 See for example Hsu (2004); Kaplan/ Schoar (2005); Chemmanur et al. (2008); Bottazi et al. (2007); Bengtson/ Sensoy (2008).
41 See Bengtson/ Sensoy (2008); Morten (2007).
42 Gompers (1995, 1996) find that young VCs exit their portfolio companies earlier and more often via an IPO than older VCs. This phenomenon is named "grandstanding" and can be motivated by the young VC's ambitions to gain additional reputation for further fundraising.

the company's agency costs decrease, the expected return rises and it is more likely the portfolio company can be made public, which leads to an increase in marginal benefits. For an experienced VC, supporting is less costly because of its superior abilities. This leads to higher monitoring and support activity with the same costs and higher benefits; or with the same benefit and lower costs for support and monitoring activity compared to inexperienced VCs.

For an experienced VC, the marginal costs of involvement decreases. When assuming a constant investment period, experienced VCs will put more effort in value-adding activities than less experienced VCs at the same cost.

Investment experience in a particular industry sector will, over time, increase VC firms' capabilities to support portfolio companies with extended contact networks or sector specific competence. Empirical evidence shows experienced VC firms in a specific industry add more value than inexperienced VCs to their portfolio companies.[43]

This leads to the following hypotheses. As monitoring and support is less costly to an experienced VC, the arranged interaction can be more flexible and the availability of VC is higher. As high availability increases the expected revenue, an experienced VC will be more available than an inexperienced one as it is less costly.

***Hypothesis 5**   VC availability will increase with an increase in the VC's experience.*

---

43   See Rosenstein et al. (1993), Sapienza et al. (1996) and Manigart et al. (2002).

Also, the costs of giving advice and personal communication with the portfolio company are lower with a more experienced VC.

> *Hypothesis 6* *VC advice activity will increase with an increase in the VC's experience.*

> *Hypothesis 7* *The frequency of personal communication between the portfolio company and the VC will increase with an increase in the VC's experience.*

Formal monitoring is an instrument to address possible agency conflicts within an investment. As experienced VCs are supposed to invest in companies with less uncertainty over future revenues and less information asymmetry, agency problems are lower.[44] Besides this, experienced VCs tend to support their investments more actively and are more frequently in personal contact with their portfolio companies, which also reduces information asymmetries. Therefore, the surplus in investing in formal monitoring activity is comparatively low. The marginal benefit of formal monitoring should be lower for experienced VCs.

> *Hypothesis 8* *The frequency of formal reporting will decrease with an increase in the VC's experience.*

---

44 See Bengtson/ Sensoy (2008).

2.3 Methodology and Results

2.3.1 Data Set

2.3.1.1 Data Description

The analysed data set of 79 British and German based VC financed biotechnology companies is a sub-sample of 280 companies from the red biotechnology sector interviewed in 2005.[45] The base sample includes companies from several biotechnology databases as well as suitable companies identified by further internet research. The population consisted of 689 companies based in Germany and the UK that were founded before 2005. The selected firms were asked to participate in a personal interview conducted by professional interviewers from a commercial market institute. The interviews with top-level firm managers took place between May and October 2005. With a response rate of 40% the sample can be considered as representative for German and British biotechnology companies in the red biotechnology sector.[46] The analysed-sub sample of 79 companies consists of companies that have at least successfully completed one venture capital financing round and have given evidence on communication frequency as well as on support activity of their VC investor. The survey data has been complemented by data from VentureXpert, a Thomson Reuters database. This additional data contained information on characteristics of the respective lead VC firm and was matched with the original data set.

---

[45] According to the OECD (2005) definition red biotechnology involves health and bioinformatics.
[46] For a detailed description on the characteristics and representativeness of the responding firms, please refer to Haagen et al. (2007).

2.3.1.2 Definition of Variables

Dependent Variables

Some of the dependent variables are interval scaled, and some are ordinal. To adjust the scales of the dependent variables, interval variables have been transformed into ordinal scales.

The dependent variables measuring the VC's monitoring activity are *formal_reporting* and *personal_communication*.

Formal_reporting measures the frequency of which a portfolio company has to provide written reports to the VC. In the questionnaire the frequency of formal reporting was measured in weeks. This information has been recoded to an ordinal scale with five specifications, according to the distribution of the answers. The scale ranges from very infrequent (1) to very often (5).

Personal_communication is a measure for the frequency of personal meetings and phone calls between the senior management of the portfolio company and the lead VC. In the questionnaire the frequency was measured in days, which has been recoded to an ordinal scale with five specifications according to the distribution of the answers. The scale ranges from very infrequent (1) to very often (5).

The dependent variables measuring VC's support activity are *availability* and *advice*.

The dependent variable *availability* measures the period of time it takes for an unexpected urgent decision that has to be made by the portfolio company, which needs the involvement of the VC. The question could be answered on an ordinal scale from longer than one week (1) to within a few hours (4).

The variable *advice* is generated by concentrating the VCs involvement in the areas of corporate strategies, organisational structures, commercialization strategy, recruiting key employees, providing contacts to cooperation partners and customers and assisting in further fundraising. The range of the variable is between no overall involvement at all (1) to very strong overall involvement (5). To form the concentrated variable, a score has been created for every observation for which there is a response to at least one item. The score is divided by the number of items over which the sum is calculated. The correlation of the advice-scale with the underlying factors is 0.89 (Cronbach's alpha), which is considered as sufficient and means the data is suitable for being indexed.[47]

Independent Variables

The institutional framework is measured by a binary coded variable *german_company*. *German_company* equals one if the portfolio company is located in Germany, representing a bank-based economic environment and equals zero if the company is located in the UK, representing a market-based economic environment.

VC experience in this study is estimated by several proxies. The experience of the VC is measured by the age of the VC firm, by the amount of funds under management,[48] as well as whether the last financing round is syndicated. These proxies are common in related literature.[49] Furthermore, experience is also measured by the biotech focus of the respective VC fund.

The binary coded variable *age_vc* equals one if the age of the lead VC is greater than 6 years at the time of the last financing

---

47 See Nunnally/ Bernstein (1994).
48 These Proxies have been suggested by Gompers (1996) and Gompers/ Lerner (1999).
49 See Bengtson/ Sensoy (2008); Kaplan/ Schoar (2005).

round and zero otherwise. This classification is in accordance to related literature analysing grandstanding-behaviour of young VCs. Within his "grandstanding-hypothesis", Gompers[50] classifies VCs under six years old as young, inexperienced VCs. *Age_vc* in the descriptive data analysis quantifies VC's age in years since founding date. For the multivariate analysis *age_vc* is implemented as a dummy variable.

VC_capital_under_management measures the capital under management of the lead VC at the time of the last investment round, in millions of Euros.

Syndication is a proxy for the aggregated level of VC experience. In a syndicated investment at least two VC firms screen investment opportunities and invest together in a portfolio company. Besides the diversification of firm-specific risk, syndication is an instrument that gets an additional opinion on investment opportunities, as well as on monitoring and support activity which limits the danger of funding "bad deals" and expands the support ability of the VCs.[51] Therefore syndication facilitates VC's experience, since overall experience in a syndicate is a bundle of two or more VCs during the period of cooperative investment. It has been shown that in syndicated investments the lead VC will spend about an hour more with the entrepreneur.[52] There is also evidence indicating syndicated investments are more likely to perform better in terms of achieving higher rates of return.[53] *Syndicated_investment* is a binary variable and equals one if capital to the entrepreneurial com-

---

50 See Gompers (1995, 1996).
51 See Lerner (1994); Gompers (1995); Gompers/ Lerner (1999); Lockett/ Wright (2001); Casamatta/ Haritchabalet (2003).
52 See Cumming/ Johan (2007); Lerner (1994); Lockett/ Wright (2001).
53 See Cumming/ Walz (2004).

pany was provided by more than one VC, and equals zero otherwise.

Another proxy for VC experience is the biotech investment focus of the lead investor. If VC's knowledge is bundled into one sector, the VC's experience in that sector is assumed to be higher since VC builds up a better understanding and thereby achieves a competitive advantage of "hard to imitate" internal resources.[54] The binary coded variable *biotechnology_focused_vc* equals one if the lead VC has an investment focus in biotechnology, and zero otherwise.

2.3.1.3 Control Variables

There might be a selection bias whereby more experienced VCs monitor less and support more, because they invest in higher-quality companies with fewer agency problems.[55] Due to the limited investment period in VC funds, VC prefers to invest in portfolio companies that look most attractive and show a lower level of uncertainty. When determining their support and monitoring activity, VCs weigh potential agency costs against potential monitoring and supporting costs. When agency problems are not as severe, support and monitoring are less likely to increase with the VC's experience. This is particularly the case with more mature portfolio companies and companies that raise more round financing.[56]

In contrast to the argument that VCs get more involved in more mature portfolio companies, Lerner[57] asserts that VCs put more ef-

---

54 See Gupta/ Sapienza (1992); Manigart (1994).
55 See Sapienza/ Gupta (1994).
56 See Bengtsson/ Sensoy (2008).
57 See Lerner (1995).

fort in portfolio companies when the need for monitoring and support is greatest and agency problems are most severe.

To address these issues, proxies for estimating the risk of the company are included. The age of the portfolio company, the stage of development of the core product and the number of previous closed VC financing rounds control for the significance of agency problems and the quality of the company. *Age_company* measures the years since company founding until the date the company has received its last financing round previous to the study date. *Later_stage* is a binary variable and equals one if the development of the portfolio company's core product is at least in a clinical stage and zero otherwise. *Financing_rounds* quantifies the number of closed financing rounds prior to the last financing round.

To test whether the absolute size of the respective financing round influences monitoring and support activity of the VC, the total amount of the last financing round is included in the model. The relative costs of monitoring and support decrease with an increase in the amount of financing in the portfolio company, therefore the VC's involvement should increase with an increasing investment size. *Amount_last_financing_round* is a measure for the amount VC invested in the last financing round in Euros.

Another control variable, which might indicate a higher level of monitoring and support, is the target exit mode. In the case where the preferred exit is an IPO, potential returns are higher and the incentive for value adding is increasing. The binary variable *IPO* equals one if the VC's aimed exit mode is an IPO and zero if the preferred exit mode is a trade sale.

I also account for the implementation of milestone agreements within a financing round. In a milestone agreement the capital infusions are paid in several stages depending on achieving previously determined benchmarks. Sahlman[58] states that staged capital is the most potent control mechanism a VC can employ. Before a fresh capital infusion takes place, prospects of the company are revaluated, within the milestone agreement. Therefore, it is expected that with a milestone agreement the monitoring frequency is decreasing. The binary coded variable *milestone_agreement* equals one if the last financing round's contract contains a milestone agreement, and zero otherwise.

A biotechnology related control variable is the dummy variable *therapeutic*, which equals one if the product development focuses on the therapeutic sector, and zero otherwise. Companies developing therapeutic medication offer higher potential gains compared to other biotechnology products, but on the other hand the development of therapeutic medication is a matter of high uncertainty associated with longer development cycles.[59] Consequently, VC might have more incentive to monitor; support activity should increase in case of a therapeutic medication development. Table 2.1 shows a summary of the independent variables and their expected influence on monitoring and support.

2.3.1.4 Summary Statistics

Table 2.2 provides a summary of the characteristics of the data. The first column concerns the whole dataset consisting of 79 VC financed portfolio companies. The statistic shows the mean for all variables used, as well as the standard deviation and minimum, as well as maximum, value for metric variables. The second column presents the characteristics for the sub-sample of 49 German companies, while the third column shows the summary statistics of 30

---

58 See Sahlman (1990).
59 See Casper (2000).

UK based biotechnology firms. The differences in the means in the German and UK sub-samples have been tested by a two-sample test of equal means.

Table 2.3 contains the pairwise correlations for all variables used. For nominal variables, Cramer's correlation coefficients have been estimated; for ordinal and metric data, Spearman coefficients.

Table 2.1
Overview of Independet Variables and Expected Impact on Monitoring and Support

The following table shows the definition of the independent variables used in the ordinal regression models as well as the expected impact on monitoring and support, whereas a '+' implies a positive and '-' a negative impact.

Variable	Variable Definition	Monitoring		Support	
		Personal	Formal	Advice	Avail.
Institutional Framework					
German company	Dummy for German based firms	-	+	-	-
VC-Experience					
Age VC	Years since foundation of VC firm	+	-	+	+
Syndicated investment	Dummy for VC financing provided by more than one investor	+	-	+	+
VC capital under management	Capital under management of the VC firm in EUR	+	-	+	+
Biotechnology focused VC	Dummy for VC funds that focus on biotech/ health care sector	+	-	+	+
Control Variables					
Amount last financing round	Amount of last financing round in EUR				
Age company	Years since foundation of portfolio company				
Financing rounds	Numbers of closed financing rounds of the portfolio company				
Later Stage	Dummy for most important product beeing in a later development stage				
Milestone agreement	Dummy for milestone agreement in the last financing round				
IPO	Dummy for the VC aiming for an IPO				
Therapeutic	Dummy for therapeutic product development				

Table 2.1: Overview of Independent Variables and Expected Impact on Monitoring and Support

Table 2.2
Summary Statistics

The following table presents the summary statistics for 79 VC financed British and German biotechnology companies. Asteriks in the last column indicate two-sample test of equal means as being significant at 1%***, 5%**, and 10%* levels.

	Total				German		UK		
N	79				49		30		
	Mean	S.D.	Min	Max	Mean	S.D.	Mean	S.D.	
Monitoring									
Formal reporting	3.1	1.1	1.0	5.0	3.4	1.0	2.8	1.3	***
Personal communication	3.2	1.3	1.0	5.0	3.1	0.2	3.3	0.3	
Support									
Availability	2.6	0.9	1.0	4.0	2.5	1.0	2.8	0.9	
Advice	3.2	1.4	1.0	5.0	2.9	0.2	3.8	1.2	***
VC-Experience									
Age VC (yrs)	18.7	17.3	2.0	60.0	18.6	18.7	19.0	15.1	
Syndicated investment (0/1)	0.7	-	-	-	0.8	-	0.6	-	
VC capital under management (bn €)	1.1	2.1	0.0	11.0	1.1	1.9	1.1	2.4	
Biotechnology focused VC (0/1)	0.4	-	-	-	0.4	-	0.4	-	
Control Variables									
Amount last financing round (mil €)	7.1	9.6	0.0	35.0	7.4	9.5	6.7	10.0	
Age company (yrs)	6.9	4.0	1.0	20.0	6.7	3.3	7.2	5.0	
Financing rounds	2.2	1.1	1.0	6.0	2.3	1.1	2.1	1.0	
Later Stage (0/1)	0.6	-	-	-	0.7	-	0.6	-	
Milestone agreement (0/1)	0.5	-	-	-	0.5	-	0.4	-	
IPO (0/1)	0.3	-	-	-	0.2	-	0.4	-	
Therapeutic (0/1)	0.4	-	-	-	0.4	-	0.4	-	

Table 2.2: Summary Statistics

Monitoring

The mean frequency of formal reporting is 3.1 (the ordinal class (2) stands for every third month up to once a year, (3) stands for every second month and the ordinal class (4) for once a month). There are country specific differences in the frequency of formal reporting, as portfolio companies based in Germany have to write a formal report more frequently than their UK counterparts. The frequency of formal reporting is positive correlated with the portfolio company's firm base in Germany. This already points in the direction of hypothesis 4.

The mean frequency of personal communication is 3.2 - (3) stands for personal communication every 61 to 90 days, (4) covers personal communication every 31 to 60 days. The frequency of personal communication is in line with a survey of Gorman and Sahlman (1998), where lead-VCs have been asked how often they pay direct attention to their portfolio companies. The frequency of personal and formal reporting is positively correlated. This indicates that some VCs are, in general, in contact with their portfolio companies more often than others.

Support

The mean of availability is 2.6, (2) stands for a decision from the VC being made within one week, (3) for a decision within a few days.

The mean of the consolidated variable advice is 3.2, whereby (3) stands for neither a weak, nor a strong overall involvement of the VC. The advice intensity differs between German and UK portfolio companies. On average VC's involvement in UK based companies is stronger than their involvement in German based portfolio companies. In addition to that, VC's advice involvement is negatively correlated with portfolio company's location in Germany, which also would be in line with hypothesis 2.

VC Experience

The mean age of the leading VC firm is 18.7 years with a standard deviation of 17 years. This indicates a high level of experience by the lead VCs. One reason for the high mean age might be the focus on lead investors, who tend to be the oldest and most experienced of the syndicate (70% of the overall investments are syndicated). The level of syndication in Germany is 80%, which is about one third higher than the syndication level of VCs investing in UK based companies. This proportion is surprising as the UK is considered to be the more experienced market for VC.

The average size of VC's total funds under management at the time of the last financing round is about 1.1 billion EUR, with a standard deviation of 2.1 billion EUR. This shows a large variation in VC firm sizes.

Table 2.3
Pairwaise Correlation

This table shows the Cramer's correlation coefficients for nominal and Spearman coefficients for ordinal and metric variables for 79 VC financed biotechnoly companies. Asteriks indicate variables as being significant at 1%***, 5%**, and 10%* levels.

		(1)	(2)	(3)	(4)	(5)	(6)	(7)	(8)
(1)	German company	1							
(2)	Formal reporting	0.26***	1						
(3)	Personal comm.	-0.11	0.34***	1					
(4)	Availability	-0.12	-0.11	0.29***	1				
(5)	Advice	-0.29***	-0.05	0.16	0.23***	1			
(6)	Age VC	-0.08	-0.03	-0.01	-0.10	0.03	1		
(7)	Synd. investment	0.13	0.15	0.05	0.11	0.09	0.11	1	
(8)	VC capital under mmgt	0.14	0.01	0.00	-0.20***	0.03	0.50***	0.27**	1
(9)	Biotech focused VC	0.04	0.16	0.15	0.26***	0.04	-0.37***	0.00	-0.01
(10)	Amount last fin. round	0.16	0.07	-0.01	-0.16	-0.07	0.35***	0.57***	0.4***
(11)	Age company	-0.01	-0.21*	-0.33	-0.11	0.00	0.30***	0.15	0.48***
(12)	Financing rounds	0.10	0.12	0.19*	-0.17	-0.08	-0.01	0.23**	0.09
(13)	Later Stage	0.05	0.02	-0.28***	-0.15	0.04	0.23**	0.09	0.19*
(14)	Therapeutic	-0.03	0.05	-0.08	-0.07	0.11	-0.06	-0.07	-0.04
(15)	Milestone agreement	0.07	0.11	-0.06	-0.10	0.01	-0.12	-0.11	-0.05
(16)	IPO	-0.13	-0.18	-0.09	-0.04	0.10	0.20*	-0.02	0.32**

		(9)	(10)	(11)	(12)	(13)	(14)	(15)
(10)	Amount last fin. round	0.00	1					
(11)	Age company	-0.21*	0.20*	1.00				
(12)	Financing rounds	0.10	0.22*	0.18	1.00			
(13)	Later Stage	-0.09	0.13	0.40***	-0.21*	1.00		
(14)	Therapeutic	0.01	-0.16	0.00	-0.24*	0.11	1.00	
(15)	Milestone agreement	0.06	-0.03	-0.09	-0.14	0.00	-0.02	1.00
(16)	IPO	-0.06	0.11	0.20*	-0.04	0.03	-0.10	-0.23

Table 2.3: Pairwaise Correlation

Control Variables

The mean financing amount in the latest VC financing round is 7.1 million EUR, a standard deviation of 9.6 million EUR indicates that a large variation in the financing amount exists. The average financing amount in Germany is 7.4 million EUR, and therefore larger than the investment amount of UK based companies, whereby differences in the means are insignificant.

The mean age of the VC financed biotechnology companies is 6.9 years since founding. As the development of a market-ready product in the biotech sector takes 12 – 15 years it makes sense that 40% of the companies in the sample are still in an early stage of development. Surprisingly, the average age of UK companies is higher, but 10% less of companies are already in a later stage of product development. 50% of all financing rounds include a milestone agreement, which means the company receives the whole amount in several stages.

According to the portfolio companies, about one quarter of lead VCs financing a German biotechnological company prefer to exit the investment with an IPO, whereas more than a third of VCs tend to exit a UK company with an IPO. This might be an indicator of the more market orientated institutional framework in the UK.

2.3.2 Methodology and Empirical Results

2.3.2.1 Ordinal Regression Model

The impact of a portfolio company's institutional framework and VC's experience on monitoring and support activity is tested using an ordinal regression model.[60] Since the dependent variable advice

---

60 For a detailed discussion of ordinal regression models see Long (1997) or Long/ Freese (2006).

is ordinal scaled and all other dependent variables are scaled in unequal time intervals, I use an ordinal scale adjusting the parameters used for comparability reasons.

After transforming the data, all dependent variables are ordinal scaled, the categories can be ordered and distances between the categories are irregular. The ordinal regression model is nonlinear and the magnitude of the change in the outcome probability, for a given change in one of the independent variables, depends on the levels of all independent variables. Therefore, the value of the estimated coefficients cannot be easily interpreted, but the prefix indicates the direction of the relation between the dependent and independent variables.

The structure of the ordinal model is

(1) $y_i^* = x_i \beta + \varepsilon_i$, where y^* is a latent variable ranging from $-\infty$ to ∞, i is the observation and ε is a random error. y^* is divided into J ordinal categories

(2) $y_i = m$ if $\tau_{m-1} \leq y_i^* < \tau_m$ for $m = 1$ to J.

The cutpoints τ_1 to τ_{J-1} are estimated and it is further assumed that $\tau_0 = -\infty$ and $\tau_J = \infty$. When the latent variable y^* crosses a cutpoint, the observed category changes. The formula for predicting probabilities in the ordinal regression model for observing $y = m$ for a given value of x is

(3) $\Pr(y = m | x) = F(\tau_m - x\beta) - F(\tau_{m-1} - x\beta)$.

In an ordinal probit model, F is the cumulative distribution function for ε with $\text{Var}(\varepsilon) = 1$.

Within the following analysis the set of exogenous variables is constant. As provided in table 2.3, the pairwise correlations of the independent variables show some high correlations that might imply multicollinearity and endogeneity issues. To address possible multicollinearity and endogeneity among the independent variables, the data is tested with the variance inflation factor (VIF). The VIF is a method of detecting the severity of multicollinearity issues. An index measures how much the variance of a coefficient is increasing because of collinearity. A common cut-off criteria for excluding a variable is a VIF > 4. Thereby the square root of the variance inflation factor tells us how much larger the standard error is, compared with what it would be if that variable were uncorrelated with the other independent variables in the equation.[61] The VIFs for all variables in the model are below 2, therefore no variables have to be excluded for reasons of multicollinearity.

2.3.2.2 Influence of Institutional Framework and Venture Capitalist's Experience on Monitoring

Formal Reporting

Table 2.4 presents the results of three ordinal regression models testing the impact of the institutional framework and the VC's experience on the frequency of formal reporting. In model III the impact of the whole set of explanatory variables is estimated to account for specification bias and collinearity. Models I and II test whether the empirical effects are robust. As assumed in hypothesis 4, the frequency of formal reporting is decreasing with market orientation of the portfolio company's head office. This finding is robust and significant for all tested models. Furthermore no other exogenous variables show a significant effect. There is no support for the impact of VC's experience on the frequency of formal reporting.

---

61 See Craney/ Surles (2002).

Personal Communication

Table 2.5 presents the results for the ordinal regression analysis of the impact of the institutional framework and VC's experience on the frequency of personal communication. The model supports hypothesis 7, as the frequency of personal communication increases with a VC older than six years and in addition, VC's funds under management are positive correlated with the level of personal communication. This effect does not hold in model I. The personal reporting frequency increases with a decrease in the portfolio company's age and an increase in previous financing rounds. The negative relation between personal communication frequency and the age of the portfolio company is in line with the assumption of Lerner[62], which states that monitoring activity is increasing with an increasing level of information asymmetry and uncertainty. This result is in contrast with the significant positive effect of the number of closed financing rounds on the frequency of personal reporting. The relation would support the Sapienza and Gupta[63] assumption that VCs become more involved in the most promising portfolio companies that feature less risk in expected gains.

2.3.2.3 Influence of Institutional Framework and Venture Capitalist's Experience on Support

Availability

Table 2.6 presents the influence of the institutional framework and VC experience on the availability of the VC. The institutional framework has no significant influence on the availability of the VC. There is ambiguous evidence for hypothesis 5. In line with hypothesis 5 is the positive impact of a VC older than six, syndicated

---

62 See Lerner (1995).
63 See Sapienza/ Gupta (1992).

investments and VC's focusing in biotech on the VC's availability. The influence of VC's capital under management is not in the expected direction, since with an increase in the firm's capital under management availability is decreasing. This may be due to reputational reasons. If VC firm's size is below a certain threshold the firm needs a disproportionate increase in reputation to signal good quality for the next fundraising cycle.[64] Therefore the availability of small sized firms might be above an optimal level where the additional costs of support are not compensated by the additional value-adding surplus of high availability when exiting the portfolio company. Another reason could be the non-linear correlation between funds under management and support activity. Kannianen and Keuschnigg (2004) and Cumming (2006) find evidence for a decreasing support activity, as above a certain threshold of firm size the experienced VC's managerial resources might be scarce. The impact of the proxies measuring VC experience is robust throughout all three models.

Furthermore, the number of previous closed financing rounds, as well as product development in a later stage, has a negative impact on VC's availability. This is in line with principal agent theory, since with an early stage of product development and only a few previous closed financing rounds information asymmetry and uncertainty is high. In this scenario it is important for the success of the entrepreneurial company that the lead VC is highly available in cases when urgent decisions have to be made. Also, the amount of invested capital in the last financing round has a negative impact on VC availability. This could be explained by considering the limited time of a VC managing their portfolio companies. It can be assumed that the time spent supervising a portfolio company is pro-

---

64 See Gompers (1995, 1996).

portional to the amount of invested money.[65] Since being highly available is time intensive for the VC, availability should increase with an increase in the amount invested by the VC.

Advice

Table 2.7 provides the estimation of the impact of the institutional framework and VC experience on VC's advice activity. With a market orientated institutional framework, the involvement of the lead VC is increasing, which is in line with hypothesis 2. Hypothesis 6 is supported by the positive relation between syndication and VC's advice activity. All coefficients are significant and robust throughout model I to III.

---

65 See Zider (1998).

Table 2.4
Ordinal Regression Model: Impact of Institutional Framework and VC Experience on Formal Reporting

This table shows the estimates of the ordinal regression model of the impact of the the institutional framework and VC experience on the frequency of formal reporting. Cutpoints are the threshold on the latent variable that result in a change in the category of the independent variable. Standard errors are in paranthesis. (Pseudo) R^2 measures the improvement in fit of the model that is due to the independent variables. The Wald χ^2 - statistic tests the hypothesis whether all coefficients jointly equal zero. Asteriks indicate variables as being significant at 1%***, 5%**, and 10%* levels.

Formal reporting	Model I Coef.	Model I se	Model II Coef.	Model II se	Model III Coef.	Model III se
Age VC	-0.29	(0.290)	-0.273	(0.290)	-0.265	(0.292)
Syndicated investment	0.206	(0.321)	0.204	(0.334)	0.277	(0.323)
VC capital under mgmt.			0.022	(0.059)	0.030	(0.065)
Biotech focused VC	0.222	(0.257)	0.208	(0.263)	0.191	(0.269)
German company	0.591	(0.270) **	0.560	(0.261) **	0.576	(0.267) **
Amount last fin. round					-0.034	(0.090)
Age company	-0.398	(0.363)	-0.352	(0.361)	-0.383	(0.379)
Financing rounds	0.205	(0.122)	0.154	(0.120)	0.188	(0.125)
Later stage			0.147	(0.273)	0.160	(0.272)
Therapeutic					0.231	(0.283)
Milestone agreement			0.231	(0.248)	0.296	(0.256)
IPO			-0.314	(0.309)	-0.272	(0.326)
Cut1	-1.287	(0.704)	-1.136	(0.743)	-1.211	(0.866)
Cut2	-0.268	(0.687)	-0.080	(0.727)	0.153	(0.903)
Cut3	-0.052	(0.682)	0.138	(0.720)	0.673	(0.812)
Cut4	1.777	(0.723)	1.978	(0.781)	1.926	(0.827)
N	79		79		79	
(Pseudo) R^2	0.06		0.073		0.074	
Wald χ^2	14.76		20.41		25.31	
$p > \chi^2$	0.022		0.026		0.013	

Table 2.4: Impact of Institutional Framework and VC Experience on Formal Reporting

Table 2.5
Ordinal Regression Model: Impact of Institutional Framework and VC Experience on Personal Communication

This table shows the estimates of the ordinal regression model of the impact of the the institutional framework and VC experience on the frequency of personal communication. Cutpoints are the threshold on the latent variable that result in a change in the category of the independent variable. Standard errors are in paranthesis. (Pseudo) R^2 measures the improvement in fit of the model that is due to the independent variables. The Wald χ^2 -statistic tests the hypothesis whether all coefficients jointly equal zero. Asteriks indicate variables as being significant at 1%***, 5%**, and 10%* levels.

Personal Communication	Model I		Model II		Model III	
	Coef.	se	Coef.	se	Coef.	se
Age VC	0.537	(0.277) *	0.531	(0.277) *	0.526	(0.284) *
Syndicated investment			0.240	(0.302)	0.230	(0.293)
VC capital under mgmt.	-0.08	(0.066)	0.117	(0.065) *	0.117	(0.066) *
Biotech focused VC	0.258	(0.243)	0.261	(0.242)	0.236	(0.239)
German company			-0.380	(0.242) **	-0.379	(0.272) **
Amount last fin. round			-0.081	(0.081)	-0.066	(0.086)
Age company	-1.102	(0.332) ***	-1.092	(0.345) ***	-1.142	(0.354) ***
Financing rounds	0.273	(0.120) **	0.290	(0.129) **	0.288	(0.139) **
Later stage	-0.307	(0.316)	-0.260	(0.327)	-0.303	(0.338)
Therapeutic					-0.022	(0.285)
Milestone agreement					-0.021	(0.247)
IPO			-0.220	(0.314)	-0.225	(0.336)
Cut1	-1.827	(0.554)	-2.288	(0.825)	-2.316	(0.878)
Cut2	-1.204	(0.552)	-1.657	(0.812)	-1.685	(0.872)
Cut3	-0.657	(0.539)	-1.095	(0.801)	-1.123	(0.867)
Cut4	0.572	(0.526)	0.170	(0.781)	0.142	(0.861)
N	79		79		79	
(Pseudo) R^2	0.107		0.119		0.119	
Wald χ^2	21.35		28.36		31.25	
$p > \chi^2$	0.002		0.002		0.002	

Table 2.5: Impact of Institutional Framework and Experience on Personal Communication

Table 2.6
Ordinal Regression Model: Impact of Institutional Framework and VC Experience on Availability

This table shows the estimates of the ordinal regression model of the impact of the the institutional framework and VC experience on the level of availability. Cutpoints are the threshold on the latent variable that result in a change in the category of the independent variable. Standard errors are in paranthesis. (Pseudo) R^2 measures the improvement in fit of the model that is due to the independent variables. The Wald χ^2 -statistic tests the hypothesis whether all coefficients jointly equal zero. Asteriks indicate variables as being significant at 1%***, 5%**, and 10%* levels.

Availability	Model I Coef.	se	Model II Coef.	se	Model III Coef.	se
Age VC	0.945	(0.341) ***	0.948	(0.350) ***	0.916	(0.353) ***
Syndicated investment	1.153	(0.376) ***	1.113	(0.382) ***	1.131	(0.401) ***
VC capital under mgmt.	-0.149	(0.061) **	-0.192	(0.071) ***	-0.186	(0.074) **
Biotech focused VC	0.799	(0.265) ***	0.932	(0.281) ***	0.961	(0.283) ***
German company			-0.077	(0.255)	-0.089	(0.253)
Amount last fin. round	-0.203	(0.097) **	-0.176	(0.101) *	-0.195	(0.109) *
Age company			0.393	(0.313)	0.461	(0.343)
Financing rounds	-0.303	(0.117) ***	-0.382	(0.126) ***	-0.408	(0.134) ***
Later stage	-0.482	(0.279) *	-0.635	(0.310) **	-0.625	(0.293) **
Therapeutic					-0.371	(0.321)
Milestone agreement			-0.349	(0.257)	-0.398	(0.258)
IPO					-0.070	(0.306)
Cut1	-3.229	(0.676)	-3.051	(0.858)	-3.403	(0.951)
Cut2	-1.411	(0.682)	-1.153	(0.869)	-1.496	(0.935)
Cut3	-0.571	(0.666)	-0.304	(0.874)	-0.633	(0.933)
N	79		79		79	
(Pseudo) R^2	0.164		0.181		0.189	
Wald χ^2	35.97		35.24		35.25	
$p > \chi^2$	0.000		0.000		0.000	

Table 2.6: Impact of Institutional Framework and VC Experience on Availability

Table 2.7
Ordinal Regression Model: Impact of Institutional Framework and VC Experience on Advice

This table shows the estimates of the ordinal regression model of the impact of the the institutional framework and VC experience on the level of advice. Cutpoints are the threshold on the latent variable that result in a change in the category of the independent variable. Standard errors are in paranthesis. (Pseudo) R^2 measures the improvement in fit of the model that is due to the independent variables. The Wald χ^2-statistic tests the hypothesis whether all coefficients jointly equal zero. Asteriks indicate variables as being significant at 1%***, 5%**, and 10%* levels.

Advice	Model I		Model II		Model III	
	Coef.	se	Coef.	se	Coef.	se
Age VC			0.060	(0.318)	0.048	(0.329)
Syndicated investment	0.547	(0.293) **	0.547	(0.293) **	0.553	(0.301) **
VC capital under mgmt.			-0.005	(0.076)	0.005	(0.075)
Biotech focused VC			0.125	(0.269)	0.117	(0.274)
German company	-0.588	(0.252) **	-0.595	(0.264) **	-0.596	(0.264) **
Amount last fin. round	-0.088	(0.078)	-0.087	(0.083)	-0.088	(0.097)
Age company					-0.062	(0.360)
Financing rounds					-0.008	(0.130)
Later stage					0.074	(0.301)
Therapeutic	0.297	(0.267)	0.298	(0.270)	0.285	(0.293)
Milestone agreement	0.243	(0.243)	0.243	(0.239)	0.237	(0.241)
IPO	0.475	(0.293)	0.472	(0.335)	0.479	(0.293)
Cut1	-1.423	(0.657)	-1.318	(0.719)	-1.410	(0.898)
Cut2	-0.753	(0.635)	-0.647	(0.702)	-0.741	(0.891)
Cut3	-0.284	(0.626)	-0.178	(0.692)	-0.271	(0.880)
Cut4	0.482	(0.629)	0.590	(0.685)	0.499	(0.874)
N	79		79		79	
(Pseudo) R^2	0.051		0.051		0.051	
Wald χ^2	17.52		18.38		22.04	
$p > \chi^2$	0.008		0.031		0.037	

Table 2.7: Impact of Institutional Framework and VC Experience on Advice

2.3.3 Discussion of Empirical Findings

The study is subject to some limitations. It is impractical to control for interdependencies among the dependent variables within the ordinal regression model setting. However, the ordinal regression model is the most appropriate choice for testing the derived hypothesis.

For measuring VC experience, I have used some proxies as experience is not immediately observable. Even though the proxies used are common in related literature and theoretically motivated, the empirical validity has not been tested. Building a "VC experience-index" containing all proxies for VC's experience was not possible, as the related variables are not highly correlated.

The proxy for the institutional framework influencing VC involvement is one-dimensional, focusing only on market orientation, as this factor is the most observed one in related literature.

The reliability of the empirical results with 79 observations is subject to the small-sample problem. With samples below 100, the small-sample behaviour of maximum likelihood estimators in ordinal models is largely unknown and the model fit is worse than with an adequate sample size. With the VIF test the small-sample problem could be alleviated, as it is more severe when the dependent variables are highly collinear.[66] Due to the small sample size, the independent variables had to be limited, therefore additional effects influencing the VC involvement, such as the equity stake of the VC or the majority of control rights could not be controlled.

Finally, the sample focuses on biotech companies, which limits the generalizability of the results on other industries.

---

66   See Long/ Freese (2006).

2.4 Conclusion

Based on a cost-benefit-analysis, the impact of the institutional framework and the VC's experience on monitoring and support activity have been analysed. The assumption was that all activities providing value to a portfolio company increase with a market orientation of the institutional environment and with an increase in VC experience, whereas activities that solely reduce agency costs are increasing with a bank-orientated financial system and a decrease in VC experience. These hypotheses have been tested in an ordinal regression model.

In line with the previously derived hypothesis is the positive impact of some of the proxies measuring VC experience on personal communication frequency, availability and the level of advice. The positive impact of a market-based financial system on advice activity and the positive relation between bank-based systems with the frequency of formal reporting are also in line with previous assumptions.

When comparing the influence of the institutional framework versus the influence of VC's experience on monitoring and support, definite driving forces cannot be identified. The frequency of formal reporting is solely determined by the institutional framework of the portfolio company's home base. The frequency of personal communication and the level of advice are determined by VC's experience. Advice activity is driven by the institutional framework as well as by VC's experience.

In addition, some control variables have a significant impact on monitoring and support. The personal communication intensity is driven by the portfolio company's age as well by the previous number of already successfully closed financing rounds. Likewise, the investment amount of the last financing round, the number of previous financing rounds and product development in a later stage influence the availability of the VC.

Since it is widely believed that monitoring and advice does add value to portfolio companies identifying the driving forces behind VC involvement sheds further light on the application of VC financing mechanism. Further research would be necessary in analysing the driving forces behind VC involvement in more detail. A study across several countries with different levels of bank- and market-based financial systems would provide more reliable results for the impact of the institutional system on monitoring and support.

Bibliography for Chapter 2

Armour, John / Cumming, Douglas (2004): The Legal Road to Replicate Silicon Valley, Working Paper University of Cambridge No. 281, March 2004.

Bascha, Andreas / Walz, Uwe (2001): Convertible Securities and Optimal Exit Decisions in Venture Capital Finance, in: Journal of Corporate Finance 7, 285-306.

Becker, Ralf / Hellmann, Thomas (2003): The Genesis of Venture Capital- Lessons from the German Experience, CESifo Working Paper No. 883, November 2002.

Bengtson, Ola / Sensoy, Berk A. (2008): Investor Abilities and Financial Contracting: Evidence from Venture Capital, Working Paper, Cornell University, August 2008.

Black, Bernard / Gilson, Ronald (1998): Venture Capital and the Structure of Capital Markets: Banks Versus Stock Markets, in: Journal of Financial Economics 47, 243-277.

Boone, Audra / Field, Laura / Karpoff, Jonathan / Raheja, Charu (2007): The Determinants of Corporate Board Size and Composition: An Empirical Analysis, in: Journal of Financial Economics 85, 66-101.

Boot, Arnoud / Thakor, Anjan (1997): Banking Scope and Financial Innovation, in: Review of Financial Studies 10, 1099-1131.

Bottazi, Laura / Da Rin, Marco / Hellmann, Thomas (2008): Who Are the Active Investors? Evidence from Venture Capital, in: Journal of Financial Economics 89, 488-512.

Campbell, Terry / Frye, Melissa (2008): Venture Capital Monitoring: Evidence from Governance Structure, Working Paper, University of Delaware, May 2008.

Casper, Steven (2000): Institutional Adaptiveness, Technology Policy, and the Diffusion of New Business Models: The Case of German Biotechnology, in: Organization Studies 21, 887-914.

Casamatta, Catherine (2003): Financing and Advising: Optimal Financial Contracts with Venture Capitalists, in: Journal of Finance 58, 2059-2086.

Casamatta, Catherine / Haritchabalet, Carole (2003): Experience, Screening and Syndication in Venture Capital Investments, in: Journal of Financial Intermediation 16, 368-398.

Chemmanur, Thomas / Krishnan, Karthik / Debarshi, Nandy (2008): How Does Venture Capital Financing Improve Efficiency in Private Firms? A Look Beneath the Surface, Working Paper, Carroll School of Management, February 2008.

Craney, Trevor / Surles, James (2002): Model-Dependent Variance Inflation Factor Cutoff Values, in: Quality Engineering 14, 391-403.

Cumming, Douglas (2006): The Determinants of Venture Capital Portfolio Size: Empirical Evidence, in: Journal of Business 79, 1083-1126.

Cumming, Douglas / Atiquah binti Johan, Sofia (2007): Advice and Monitoring in Venture Finance, in: Financial Markets and Portfolio Management 21, 3-43.

Cumming, Douglas / MacIntosh, Jeffrey (2001): Venture Capital Investment Duration in Canada and the United States, in: Journal of Multinational Management 11, 445-463.

Cumming, Douglas / Walz, Uwe (2004): Private Equity Returns and Disclosure Around the World, LSE Working Paper, April 2004.

Franzke, Stefanie / Grohs, Stefanie / Laux, Christian (2004): Initial Public Offerings and Venture Capital in Germany, in: The German Financial System, Krahnen / Schmidt (ed.), University Press, Oxford, 233-260.

Gaida, Michael (2002): Venture Capital in Deutschland und den USA, Wiesbaden.

Gilson, Ronald (2003): Engineering a Venture Capital Market: Lessons from the American Experience, in: Stanford Law Review 55, 1067-1103.

Gompers, Paul (1995): Optimal Investment, Monitoring, and Staging of Venture Capital, in: Journal of Finance 50, 1461-1489.

Gompers, Paul (1996): Grandstanding in the Venture Capital Industry, in: Journal of Financial Economics 42, 133-156.

Gompers, Paul / Lerner, Josh (1999): What Drives Venture Capital Fundraising, NBER Working Paper, January 1999.

Gompers, Paul / Lerner, Josh (2001): The Venture Capital Revolution, in: Journal of Economic Perspective 15, 145-168.

Gompers, Paul / Lerner, Josh (2006): The Venture Capital Cycle, 2nd Edition, MIT Press, Cambridge.

Gorman, Michael / Sahlman, William (1989): What Do Venture Capitalists Do?, in: Journal of Business Venturing 4, 231-248.

Haagen, Florian / Haeussler, Carolin / Harhoff, Dietmar / Murray, Gordon / Rudolph, Bernd (2007): Finding the Path to Success: The Structure and Strategies of British and German Biotechnology Companies, Anglo-German Biotech Observatory 2006 Company Survey, AGBO Report LMU Munich.

Haagen, Florian (2008): Venture Capital Financing Practices in Germany and the UK, Hamburg.

Hackethal, Andreas (2004): German Banks and Banking Structure, in: German Financial System, Krahnen / Schmidt (ed.), Oxford University Press, 69-105.

Hart, Oliver / Moore, John (1999): Foundation of Incomplete Contracts, in: Review of Economic Studies 66, 115-138.

Hege, Ulrich / Palomino, Frédéric / Schwienbacher, Armin (2003): Determinants of Venture Capital Performance: Europe and the United States, Working Paper University of Amsterdam, November 2003.

Hsu, David (2004): How Much Do Entrepreneurs Pay for Venture Capital Affiliation, in: Journal of Finance 59, 1805-1844.

Inderst, Roman / Müller, Holger (2004): The Effect of Capital Market Characteristics on the Value of Start-Up Firms, in: Journal of Financial Economics 72, 319-356.

Jeng, Leslie / Wells, Philippe (2000): The Determinants of Venture Capital Funding: An Empirical Analysis, in: Journal of Corporate Finance, 241-289.

Kanniainen, Vesa / Keuschnigg, Christian (2003): The Optimal Portfolio of Start-up Firms in Venture Capital Finance, in: Journal of Corporate Finance 9, 521-534.

Kanniainen, Vesa / Keuschnigg, Christian (2004): Start-up Investment with Scarce Venture Capital Support, in: Journal of Banking and Finance 28, 1935-1959.

Kaplan, Steven / Strömberg, Per (2003): Financial Contracting Theory Meets the Real World: An Empirical Analysis of Venture Capital Contracts, in: Review of Economic Studies 70, 281-315.

Kaplan, Steven / Strömberg, Per (2004): Characteristics, Contracts, and Actions: Evidence from Venture Capital Analysis, in: The Journal of Finance 59, 2177-2210.

Kaplan, Steven / Martel, Frederic / Strömberg, Per (2007): How Do Legal Differences and Experience Affect Financial Contracts?, in Journal of Financial Intermediation 16 (3), 273-311.

Kaplan, Steven / Schoar, Antionette (2005): Private Equity Performance: Returns, Persistence, and Capital Flows, in: Journal of Finance 60, 1791-1823.

Leleux, Benoit / Surlemont Bernard (2003): Public Versus Private Venture Capital: Seeding or Crowding Out? A Pan-European Analysis, in: Journal of Business Venturing 18, 81–104.

Lerner, Josh (1994): The Syndication of Venture Capital Investments, in: Financial Management 23, 16-27.

Lerner, Josh (1995): Venture Capitalists and the Oversight of Private Firms, in: The Journal of Finance 50, 301-318.

Lerner, Josh / Watson, Brian (2008): The Public Venture Capital Challenge: The Australian Case, in: Venture Capital 10, 1-20.

Lingelbach, David / Murray, Gordon / Gilbert, Evan (2008): The Rise and Fall of South African Venture Capital: A Coproduction Perspective, Working Paper, School of Business and Economics, University of Exeter, 2008.

Lockett, Andy / Wright, Mike (2001): The Syndication of Venture Capital Investments, in: Omega 29, 375-390.

Long, Scott (1997): Regression Models for Categorical and Limited Dependent Variables, in: Advanced Quantitative Techniques in the Social Sciences, Vol. 7, Thousand Oaks.

Long, Scott / Freese, Jeremy (2006): Regression Models for Categorical Dependent Variables Using Stata, 2nd Edition, Texas, Stata Press.

Manigart, Sophie / Baeyens, Kathleen / Van Hyfte, Wim (2002): The Survival of Venture Capital Backed Companies, in: Venture Capital 4, 103 -124.

Marti, José / Balboa, Marina (2001): Determinants of Private Equity Fund Raising in Western Europe, SSRN Working Paper, 2001.

Morten, Sorensen (2007): How Smart is Smart Money? A Two-Sided Matching Model of Venture Capital, in: Journal of Finance 62, 2725-2762.

Nunally, Jum / Bernstein, Ira (1994): Psychometric Theory, 3rd Edition, New York.

OECD (2005): A Framework of Biotechnology Statistics, Paris.

Repullo, Rafael / Suarez, Javier (2004): Venture Capital Finance: A Security Design Approach, in: Review of Finance 8, 75-108.

Rosenstein, Joseph / Bruno, Albert / Bygrave, William / Taylor, Natalie (1993): The CEO, Venture Capitalists, and the Board, in: Journal of Business Venturing 8, 99-113.

Sahlmann, William (1990): The Structure and Governance of Venture-Capital Organizations, in: Journal of Financial Economics 27, 473-521.

Sapienza, Harry (1992): When Do Venture Capitalists Add Value, in: Journal of Business Venturing 7, 9-27.

Sapienza, Harry / Gupta, Anil (1992): Impact on Agency Risks and Task Uncertainty on Venture Capitalists – CEO Interaction, in: Academy of Management 37, 1618-1632.

Sapienza, Harry / Manigart, Sophie / Vermeir, Wim (1996): Venture Capital Governance and Value Added in Four Countries, in: Journal of Business Venturing 11, 439-469.

Schefzyk, Michael (2000): Finanzieren mit Venture Capital, Stuttgart.

Schmidt, Klaus (2003): Convertible Securities and Venture Capital Finance, in: Journal of Finance 58, 1139-1166.

Schmidt, Reinhard / Tyrell, Marcel (2004): What constitutes a Financial System in General and the German System in Particular, in: The German Financial System, Krahnen / Schmidt (ed.), University Press, Oxford, 19-57.

Schwienbacher, Armin (2008): Venture Capital Investment Practices in Europe and the United States, in: Financial Markets and Portfolio Management 22, 195-217.

Stinchcombe, Arthur (1965): Social Structure and Organizations, in: J. G. March (ed.), Handbook of Organizations, Chicago, 142-193.

Tykvová, Tereza (2007): What Do Economists Tell Us About Venture Capital contracts?, in: Journal of Economic Surveys 21, 65-89.

Zider, Bob (1998): How Venture Capital Works, in: Harvard Business Review, November-December 2008, 131-139.

3 The Role of Venture Capital and Corporate Venture Capital in Financing Biotech Companies

3.1 Introduction

Innovative entrepreneurial firms with novel technologies often lack funding for the development and introduction of new products. High informational asymmetries, which are common in an R&D intensive environment, are often the reason for this. Moreover, in many instances insufficient funds are accompanied by a lack of required research resources and managerial knowledge for realizing the original research ideas, which would take years to develop independently by the entrepreneurial company. Consequently, financial prospects and survival chances are uncertain. Investing time and capital in these entrepreneurial companies is a hazardous decision.[67] Therefore, young and small research intensive companies frequently rely on specialised financial intermediaries focusing on investing in highly innovative and independent entrepreneurial companies.[68]

Strategic equity alliances with larger corporations, hereafter referred to as corporate venture capital (CVC) and venture capital (VC) meet these particular issues of innovative companies and play a dual role in financing and advising entrepreneurial companies: Besides providing equity or equity linked finance to entrepreneurial firms, VC and CVC also provide value-adding activities to their portfolio companies. Potential conflicts of interest between the in-

---

67 See Stuart et al. (1999).
68 For VC see Gorman/ Salman (1989), for CVC see Dushnitsky/ Lenox (2005).

termediary and their portfolio company resulting from high information asymmetries are mitigated by the VC's and CVC's monitoring and support function.[69]

When engaging with an entrepreneurial company VC and CVC do not only compete on the price of equity but also on the level of support they can credibly provide in the case sufficient VC and CVC funds are available. VC distinguishes from CVC by their value-adding contributions to the commercial success of their portfolio company.[70]

Previous theoretical and empirical studies used a simultaneous investment framework for studying the respective value-adding activity from VC and CVC as well as a joint analysis of the adding value contribution on firm performance. In an overall industry framework, it has been shown the value-adding contribution of VC is focused on managerial related advice, whereas CVC focuses on providing research related resources and advice when both are invested in a portfolio company at the same time. These value-adding contributions have been tested ex post, after the investment of VC and CVC has already taken place.[71] However, the demand for a specific value-adding contribution describing a portfolio company's individual need for support in a certain area accounting for the allocation of VC and CVC investments has not yet been investigated.

Hence, in this paper I will explain the preference for either CVC or VC investment from an entrepreneurial firm's perspective. If the capital allocation from a portfolio company's perspective were efficient the likelihood of receiving VC and / or CVC should be de-

---

69   See Tykvova (2003).
70   See Maula et al. (2005) and Hellmann (2002).
71   See Hellmann (2002); Maula et al. (2005) and Ivanov/ Xie (2010).

pendent on an entrepreneurial company's specific need for value-adding activities.

I utilize a dataset of European biotechnology companies containing VC as well as CVC investments. I analyse whether VC and CVC act as supplements or as complements in their role of meeting the portfolio companies' needs in advising and adding value. I find VC is more likely if the company has a general demand for managerial related advice whereas the likelihood for CVC increases with a demand for research related resources.

The study contributes to previous literature[72] insofar as I find empirical support for the assumption that entrepreneurial companies already ex ante choose their financier, VC versus CVC, based on their company specific demand for the respective value-adding activities.

The study focuses on the biopharmaceutical industry, which has become one of the most research-intensive sectors within the last years. The development of new products in that industry is very capital-intensive and takes up to 10-15 years. Compared to other industry sectors the risk of failure is higher. These characteristics make the financing of biotechnology even more challenging. The investment in entrepreneurial companies in the biopharmaceutical sector for VC, and CVC to the same extent, is of high interest.[73]

The remainder of the paper is organized as follows. Section 2 provides a brief definition of VC and CVC from an institutional and entrepreneurial perspective and discusses the economic relevance

---

72 See Maula et al. (2005) and Dushnitsky (2006).
73 For an overview on the financing of the biopharmaceutical sector please refer to the report of the European Comission (2009).

of the dual role of financing and advising portfolio companies. In section 3 the hypotheses are being derived. Section 4 provides an overview of the data used and tests the hypotheses using a seemingly bivariate probit model. After discussing the theoretical and empirical results the final section concludes the paper.

3.2 Theoretical Framework

3.2.1 Venture Capital versus Corporate Venture Capital

There is a clear distinction between VC and CVC from an institutional point of view. VC is institutional capital provided by institutional investors and wealthy individuals, acting as financial intermediates investing primarily in privately owned early stage companies, often technology related, with large growth potential. CVC investments are equity investments by established, often R&D intensive firms, in research intensive entrepreneurial companies. In the course of this study CVC is defined as equity that is provided by a strategic alliance partner within a cooperation between the portfolio company and a pharmaceutical firm.[74]

VC, as well as CVC institutions provide equity and additional value-adding services to their portfolio companies.[75] Therefore they both address financial, as well as knowledge-based issues of research-intensive entrepreneurial companies. The original purpose for financing and advising entrepreneurial companies diverges for VC and CVC. Whereas VCs pursue a solely monetary objective in engaging with their portfolio companies, CVC additionally aims for

---

[74] For a substantial academic definition of CVC please refer to Dushnitzky/ Shapira (2009).
[75] See Tykvova (2003) and Dushnitsky (2006).

strategic benefits in allying with a research-focused company to add strategic value to their own corporate parent organization.[76]

From an entrepreneurial perspective the classification of VC and CVC is more complex. From this point of view, VC firms' performance might be measured in terms of their ability to add value in addition to capital infusions. Earlier research shows VC firms play an important role in professionalizing the firms in which they invest by connecting them with potential clients, suppliers and cooperation partners, and attracting additional funding.[77] The benefits of CVC are not described in such detail as those of VC. Nevertheless literature suggests that the value-adding contributions to their portfolio companies are different from those of VC.[78] This might be due to the different objectives of VC and CVC when engaging with an entrepreneurial company.

From a portfolio company's perspective the most important value-adding services are assistance in arranging additional financing, support in management related areas, access to tangible and intangible research resources, as well as an increasing certification by an engagement with a prominent financier.[79]

According to empirical studies the main field of advice activities diverges between VC and CVC.[80] Gorman and Sahlman (1989) find that the main areas of VCs advice involve assisting with further external financing, supporting strategic decision making and recruiting key executives. For CVC the focus in value-adding activities lies in provision of tangible resources, access to intellectual property of

---

76 See Sykes (1990) and Dushnitsky (2006).
77 See Sapienza (1992); Rosenstein et al. (1993); Barney et al. (1996).
78 See Hellmann (2002) and Maula et al. (2003).
79 See Maula et al. (2005) and Gorman/ Salman (1989).
80 See Gorman/ Sahlman (1989) and Dushnitsky/ Lenox (2005).

the corporation and increasing the reputation by allying with a certified pharmaceutical firm.[81]

3.2.2 Marginal Utility of Support

In account of the institutions' dual role in financing and advising, double moral hazard problems occur in VC and CVC scenarios, as sufficient effort from each involved party is necessary for the entrepreneurial company's success.

The specification of the level of moral hazard varies since VC and CVC have different organizational and incentive structures. CVCs are structured as corporate subsidiaries and have lower incentive based compensation. This leads to the assumption CVC would add value less successfully and less efficiently to its portfolio companies.[82] This assumption contradicts empirical studies that provide evidence for CVC being at least as successful as VC in adding value to their portfolio companies.[83]

These empirical findings are also in line with a model from Casamatta.[84] The model provides an economic motivation for the VC's dual financing and advising role when engaging in a portfolio company. The dual role of financing and advising is meant to be important as efficient effort in contributing value-adding is only provided when additional equity is being invested. Since CVC also provides capital and advice the argument of the VC-model can be applied analogously in the following.[85]

---

81 See Maula et al. (2005).
82 See Gompers/ Lerner (2000).
83 See Gompers/ Lerner (2000) and Ivanov/ Xie (2010).
84 See Casamatta (2003).
85 See Aghion/ Tirole (1994) for a motivation of effort provided by strategic alliance partners.

The core assumptions of the model are that entrepreneurs endowed with creativity and technical skills necessary to develop innovative ideas, may lack business expertise and managerial skills. They are therefore reliant on advice from an external authority. Another assumption is that the effort exerted by an entrepreneur has to be more efficient than from an external authority, which includes all research projects that rely on the entrepreneur's human capital. Developing products in the biotechnology sector is a long-term process with a complex environment of legal restrictions and regulatory standards that have to be met before a product reaches the market. Even if the entrepreneur is technically skilled he may not only need support in managerial areas but also regarding the know-how in conducting research that leads to a marketable product at the end.[86] Since each VC manages a large number of portfolio companies the time supervising a company is limited. For this reason VCs are assumed to be less knowledgeable about the industry and technology than entrepreneurs.[87] Consequently, the advice of the external authority and the effort of the entrepreneur are, in reality, both needed to optimize the firm's value and for the model assumptions to hold.

Given a moral hazard framework VC and CVC, as well as the research unit, have to exert certain levels of effort, which are not observable. To induce this effort, proper incentives must be provided to both of them. In the case of VC, the research unit and the VC have to be compensated by the allocation of cash flow rights. In the case of CVC the compensation consists of cash flow and control rights.

---

86 See Stuart et al. (1999).
87 See Zider (1998).

Based on the model, the firm value is maximized when the financing partner with the highest marginal utility of effort is investing.[88] This means VC and CVC act as substitutes in cases where their support ability, and therefore efficiency, is equal and act as complements when VC ability in providing advice in certain areas is superior to CVC's ability and vice versa.

The core finding in the model shows the entrepreneurial company seeks advice from an investor rather than a pure consultant or strategic alliance partner, irrespective of the company's possible cash constraint. A dual role in financing and advising is the most efficient instrument for adding-value to an entrepreneurial company for an external authority.

Given that if capital allocation were efficient, the portfolio company would be financed by the financier that provides the most efficient advice corresponding with the company's specific demand for support. Whether VC and CVC act as substitutes or complements in financing and advising companies according to the portfolio company's specific demand for support is derived theoretically in the following section.

---

88 See Anand/ Galetovic (2000).

3.3 Supporting Entrepreneurial Companies

3.3.1 Corporate Venture Capitalist's and Venture Capitalist's Support

Management support

Professionalizing entrepreneurial companies is a core business of VC activity. Due to its specialization in monitoring and supporting companies VC has a greater and wider experience in helping entrepreneurial companies to survive and to manage early growth than CVC does.[89] Nowadays being in an experienced industry, VCs have learnt to support their portfolio companies in organizing and structuring the firm appropriately in all stages of growth. VCs have typically screened, monitored and supported numerous start-ups, from firm foundation and the initial investment to the liquidation event before investing in an additional portfolio company.[90] They play an important role in professionalizing entrepreneurial companies and give advice in all management related areas.[91] It is therefore assumed that with an increasing demand for managerial related advice of the firm, the likelihood for VC financing increases.

> *Hypothesis 1* *The likelihood for VC will increase with a need for managerial related advice.*

---

89 See Maula et al. (2005) and Dushnitsky (2006).
90 See Gorman/ Sahlman (1989).
91 See Hellmann/ Puri (2002).

Access to tangible and intangible research resources

VCs are credited with superior management and financial knowledge. Yet they are likely to have less authority than potential CVCs in areas directly related to the core business of their portfolio companies. Pharmaceutical companies have broad specialist knowledge in their sectors of competence. They are expected to have a profound and dynamic understanding of technological developments since they are spending large amounts of money on R&D and market research. They also have previously established marketing, distribution and research units. The technology related knowledge in combination with already existing customer relationships enables a more profound understanding of future market needs than that which is typically available in entrepreneurial companies developing a novel product for a yet to be established market. The resources of the parent pharmaceutical company enable CVC to aid in the growth and maturation of product development and commercialization of their portfolio companies.[92]

CVC is considered to provide more efficient advice in research related areas than VC would be able to. The depth of market understanding of pharmaceutical companies in combination with access to research resources of large pharmaceutical companies should increase the likelihood of CVC funding, if there is a demand for research related advice of the portfolio company. The demand for access to intellectual property is also assumed to increase the likelihood for CVC financing, as these resources may already exist in the parent pharmaceutical company or at least are easier to access with the backing of an already established pharmaceutical company. Another aspect of accessing research related resources is the possibility of using tangible research resources, such as the

---

[92] See Maula et al. (2005) and Ivanov/ Xie (2010).

pharmaceutical company's laboratories. The joint use of research facilities will reduce costs of the product development process. Therefore it is expected that with a demand for reducing costs, in the research process, the likelihood for CVC increases.

> *Hypothesis 2 The likelihood for CVC will increase with a demand in accessing tangible and intangible research related resources.*

i. *The likelihood for CVC will increase with an increasing need for research related advice.*

ii. *The likelihood for CVC will increase with an increasing need for access to intellectual property.*

iii. *The likelihood for CVC will increase with an increasing need for reducing costs in the research process.*

Blocking competition

Considering the VC's ability to block competition, it can be argued that large corporations are considered to be less flexible than VCs in observing innovation which does not meet their own product development. CVC runs the risk of ignoring or being unaware of novel competitive threats from high potential entrepreneurial businesses exploiting new technologies.[93] This would make pharmaceutical corporations insensitive and inappropriate partners to small innovative companies. On the other hand, VCs are specialised in focusing on the supervision of small entrepreneurial firms with high growth potential. In this role, they assist in the development of the portfolio companies' strategic perspective. This does include giving business advice and acting as a sounding board to

---

93 See Maula et al. (2005).

the firm's management.[94] VCs are more likely than CVCs to provide their portfolio companies with information concerning potentially competitive responses on their market niche when their portfolio company's growth orientated strategy is compromised.[95] This argument is even stronger in the special case when two competing companies have the same VC. In this case the VC is able to observe and discipline the non-cooperative behaviour of one party, for example absorbing R&D knowledge.[96] This leads to the assumption that VC investment becomes more likely with a demand for blocking competition to the portfolio company.

Hypothesis 3 a) The likelihood for VC will increase with a need for blocking competition.

The prospect of market entry from a highly innovative, research intensive product developed by a biotech company is very uncertain. For this reason the incentive of a pure strategic alliance partner to absorb know-how and to develop the product on its own in an early development stage is low. As soon as certain performance targets have been attained and the product has reached a later stage of development the strategic alliance partner might be more tempted to continue product development without the research unit of their portfolio company, and the potential for hold up problems increases. If the strategic alliance partner has an equity stake in the portfolio company, an incentive is implemented to protect knowledge against competitors. The pharmaceutical company's commitment on knowledge protection under hold up conditions

---

94 See Sapienza et al. (1996).
95 See Maula et al. (2005).
96 See Lindsey (2008).

becomes more credible with a dual role of financing and advising of the CVC.[97] It can be assumed that the incentive for blocking competition in later product development stages is more severe when the allied pharmaceutical firm has an equity stake in the company. Consequently, CVC investment becomes more likely when there is a high demand for blocking competition and CVC and VC can be seen as substitutes in their ability of blocking their portfolio company's competition.

> ***Hypothesis 3 b) The likelihood for CVC will increase with a need for blocking competition.***

Increasing certification

A VC is actively involved in advising and monitoring its portfolio companies by being in regular contact with the company's management.[98] Therefore, VC is perceived to be an informed agent, knowing about the real quality of the portfolio company. An ongoing investment of a VC may send a positive signal about the firm's quality and provide a certification benefit which enables the biotech firm to obtain other resources like additional financing or strategic alliances.[99] The positive signal of VC investment is considered to be reliable since VC risks harming its own reputation if false signals are sent to market participants. In this context, it has been shown that equity-based linkages between entrepreneurial companies and certified financiers, as well as linkages with strategic alliance partners, enable the companies to go public more quickly and with a higher market valuation than those with less

---

97 See Cestone/ White (2003).
98 See Hochhold (2010).
99 See Megginson/ Weiss (1991).

prominent ties.[100] As reputation is a valuable asset in various regards, a research unit is more likely to engage with a VC if the partner's reputation is of high interest.

Hypothesis 4 a) The likelihood for VC will increase with a demand for a reputational partner.

In the case of a CVC investment, the rationale for a reputed financier is equal to an investment of a reputational VC.[101] It is assumed VC and CVC can be seen as substitutes in providing a certification benefit for their portfolio companies as both institutions are considered as informed agents regarding the quality of their portfolio companies.[102] CVC, more so than VC, is accredited with an outstanding ability to attract new business partners. In biotechnology it is difficult for an outside investor to value the potential of a new technology with unsure prospects. The investment of a big pharmaceutical company may send an even more trustworthy signal of the attractive prospects of the research unit's products to less informed outsiders than VC would be able to.[103] These equity investments of pharmaceutical companies help to reduce information asymmetries about portfolio companies' real quality.

Hypothesis 4 b) The likelihood for CVC will increase with a demand for a reputational partner.

---

100 See Gompers (1996).
101 See Stuart et al. (1999); Gompers/ Lerner (2000) and Maula et al. (2005).
102 See Dushnitsky (2006).
103 See Stuart et al. (1999) and Gulati/ Higgins (2003).

3.3.2 Information Asymmetry

Young, high-technology firms in early stages of development frequently do not generate positive cash-flows from their research activities and consequently are dependent on outside financing. These companies are surrounded by uncertainty and information asymmetries. Strategic alliances often involve an equity investment in the research unit. An equity investment within a strategic alliance agreement may not only be a sign of the research unit's insufficient funds, but may also mitigate problems resulting from information asymmetries in between the involved parties, in accordance to VC mechanisms.[104] In the early development stage, the portfolio firms' bargaining power is considered to be low. Therefore in early development stages it is unlikely for these young firms, facing financing restrictions, to allocate the property rights in an efficient way according to the incomplete contracts theory.[105] Corresponding with the incomplete contracts theory, sufficient property rights should be allocated to the entrepreneurial company as long as the marginal utility of the research effort on the value of the final output is greater than the marginal utility of the financing partners' investment.[106] This means that in early stages of product development, where the effort of the entrepreneur is crucial for product success, property rights should remain with the entrepreneur in the portfolio company to induce sufficient effort. Thus it might be more efficient to induce the researcher's effort by pure financial VC engagement instead of pursuing financial and strategic interests of a CVC that transfers property rights to the pharma-

---

104 For an overview of VC mechanism mitigating moral hazard see Tykvová (2007).
105 See Lerner/ Merges (1998) and Aghion/ Tirole (1994).
106 See Aghion/ Tirole (1994).

ceutical company. In early stages of product development a VC investment induces the entrepreneur to exert sufficient effort.

Early stage research projects, particularly in biotechnology, are highly complex and uncertain, making it very difficult to contractually specify the allocation of the provided research resources, as the demand for resources is not always apparent ex ante. The specification of the contributions to the respective research project by the biotech firm is even more challenging since these firms often have additional research projects with other cooperation partners or of their own. The alliance partner cannot observe whether the biotech firm puts the contracted effort to the projects that are part of the alliance agreement.[107] Furthermore the effort carried out by the biotech firm is not contractually enforceable.[108]

In contrast to a pharmaceutical company, VC can commit ex ante to any surplus sharing rule. Moreover, the CVC can take advantage of knowledge spill overs to learn about the project and develop the innovation without the entrepreneur's participation.[109] There are also reasons why CVCs add less value to their portfolio companies than VCs do. Since CVCs usually embody structures like divisions within pharmaceutical companies they lack the monetary incentive scheme that VC managers motivate to exert effort.[110] This missing incentive scheme is more severe in early stages of product development as information asymmetries are higher.

Given this, it seems VC is more prosperous in financing companies in the early stages of product development since less contracting and monitoring problems occur. On the other hand, the chance

---

107 See Lerner/ Merges (1998).
108 See Aghion/ Tirole (1994).
109 See Anand/ Galetovic (2000).
110 See Ivanov/ Xie (2010).

of a CVC absorbing portfolio company's know-how is low, which makes CVC more likely in early stages of product development,[111] in terms of the impact of information asymmetry on the likelihood of receiving CVC

I expect that the likelihood for CVC financing increases with an increase of the portfolio company's age and with early stages of product development and decreases with a research focus of the portfolio company on therapeutics.

VC investment is more efficient with regard to the compensation of the entrepreneur, the allocation of entrepreneurial effort and the financier's institutional structure when high information asymmetries are existent. Therefore I assume the likelihood for VC will increase with a high level of information asymmetry regarding the prospects of the portfolio company. Information asymmetries are high if the portfolio company is young and a previous company track record is missing; the older the company the longer its track record. Consequently, the level of uncertainty is decreasing with past observable performance. In early stages of development the prospects of products are very unsure which leads to a high uncertainty about company development. Companies that focus on developing therapeutic medication offer, on the one hand, higher potential profits as compared to other biotechnology products; on the other hand, therapeutic medication is subject to longer development cycles.[112] The latter point increases the uncertainty of product development in early stages even more. It is therefore assumed that the likelihood for VC financing increases with a decrease of the portfolio company's age, early stages of product development and a research focus of the portfolio company on therapeutics.

---

111 See Ceston/ White (2003).
112 See Casper (2000).

>*Hypothesis 5 a) The likelihood for VC / CVC will decrease / increase with an increase in age of the portfolio company.*

>*Hypothesis 5 b) The likelihood for VC and CVC will increase with an early stage of product development.*

>*Hypothesis 5 c) The likelihood for VC / CVC will increase / decrease with a product development focus on therapeutics.*

3.4 Methodology and Results
3.4.1 Field of Study

The study focuses on the pharmaceutical sector. The analysed portfolio companies are within the biotech industry and have already entered a vertical strategic alliance with a large pharmaceutical company. CVCs investing in entrepreneurial companies not only have predominantly monetary intentions like a VC but also follow strategic objectives. They tend to create value by accessing complementary resources. R&D intensive firms need to form alliances with a partner with complementary resources to ensure a timely and successful product development and introduction.[113] A more forward-looking approach to collaborate in a strategic alliance may

---

113 See Teece (1986).

be the CVCs' intention to supplement its own internal R&D efforts. Technologies, covered by strategic alliances and CVCs can be observed in all stages of product development, from the earliest stages of research development, and almost to the final step of a regulatory approval. Unlike contractual provisions between a VC and their portfolio company which focus on the allocation of cash-flow and control-rights,[114] strategic alliances with equity involvement embody a more complex contract structure since they also provide for the allocation of property-rights.[115] Similar to a VC engagement the CVCs engagement is midterm orientated but leads ideally to an acquisition of the portfolio company. Acquisitions are a possibility for large pharmaceutical institutions to obtain valuable resources out of a vertical alliance. Within an acquisition of a former alliance partner R&D costs can be reduced and the number of potential products in the pipeline can be increased. It also has been shown that the post acquisition process with a pre-acquisition alliance relationship is less problematic in comparison to an unknown target, since the acquisition company has the possibility to gain inside information on the quality of this possible target as well as potential technology related problems. Collaboration partner specific absorption capacity and joint problem solving arrangements could also be tested in strategic alliances before acquiring the company.[116] This is why CVC is an equally important financier in the pharma-biotech sector as VC.[117] The pharmaceutical industry is most suitable for the comparison of VC versus CVC value-adding contributions.

---

[114] Control rights in the SA context are the allocation of decision rights that cannot be contractually specified ex ante due to the complexity of joint R&D projects [Lerner/ Merges (1998)].
[115] See Aghion/ Tirole (1994).
[116] See Al-Laham et al. (2010); Higgins/ Rodriguez (2006).
[117] See Dushitsky (2006).

3.4.2 Data Set

3.4.2.1 Data Collection and Description

The data has been collected in an international online survey of biotechnology firms from Austria, Belgium, Denmark, Finland, Germany, Italy, Netherlands, Sweden, Switzerland, UK, U.S., and Canada. Biotechnology companies have been identified by several databases like bioscan and biocom. The respective senior head of management or business development was asked to participate. The survey took place between July and October 2008. Up to three reminders have been administered to follow up on firms that had not replied upon a certain date. At the end of the survey period 362 senior heads participated in the survey, which represents a response rate of 16%. The survey focused on strategic alliance relations and the related allocation of control rights, it also contained general information about the firm, the current main field of activity, alliances, corporate strategy and financing.

The objective of the study is to shed light on whether the specific reasons for engaging in an alliance in general also influence the likelihood of receiving VC and CVC. I assume that the reasons to ally also discover a demand for support or rather a lack of expertise in certain areas. In some circumstances, the discovered demand may have already been satisfied by entering into a strategic alliance. To focus on money engagements and control for previous strategic alliance activity, only companies that have already entered a strategic alliance and have already received institutional capital are analysed. Given this, the empirical analysis is based on 174 firms that have already entered one or more pharmaceutical strategic alliances and have received VC and / or CVC. These firms have a similar background when an outside investor gets involved; the basic motivation for engaging in an alliance is already satisfied and additional demand for support can be revealed by alliances with an equity stake, or VC.

3.4.2.2 Definition of Variables

The dependent variables in the analysis are whether the company has received VC and / or CVC financing. The binary coded variables equal one if the biotech firm already has received VC respectively CVC, and zero otherwise.

The independent variables can be divided into three categories: Reasons for engaging in an alliance as a proxy for the demand of support, proxy variables describing the level of information asymmetry regarding each company and further control variables.

The motivation for VC and CVC investments are described by general reasons for engaging with a strategic partner. These proxies are binary scaled and equal one if the respective reason for entering a pharmaceutical alliance is considered to be important to extremely important, and equal zero otherwise. The covered reasons for engaging with a partner are the importance of access to intellectual property (*ip_access*), reducing costs for example access to partner's lab equipment (*reducing_costs*), the experience of a partner in conducting research (*exp_research*), blocking competition (*blocking_comp*), the experience of a partner in marketing / distribution (*exp_management*) and the reputation of the partner (*reputation*).

Information asymmetry is covered by proxies measuring the level of uncertainty of possible prospective company success.[118] The variable *age* measures the age of the portfolio company in years

---

[118] The used proxies of information asymmetries are common in related literature: Gompers (1996) and Lerner (1994).

since founding. The binary coded variable *early_stage* equals one if the internally developed products / technologies are in clinical phase I or in an earlier stage of development, the variable equals zero otherwise. The dummy variable *therapeutic* equals one if the company is active in the therapeutic sector, zero otherwise. As the development of products in the therapeutic sector takes longer than in other sectors the level of uncertainty regarding success is higher than in product development in the non-therapeutic sector.

3.4.2.3 Control Variables

In addition to the focus of the study comparing the value adding-ability of VC versus CVC, the study is controlled for additional impacts on the likelihood for VC and CVC financing.

The funding by external equity might be mainly due to the portfolio company's insufficient funds and not due to a support demand in certain business or research areas. Therefore there is a control on whether a main motivation receiving equity is a lack of sufficient funds. I expected that the demand for receiving money has a positive impact on CVC, as well as VC, funding.

In line with the demand for equity is the general preference of a company for payment terms over control rights. If there is a preference for giving up control rights in favour of additional payment terms, a positive impact on the likelihood of receiving VC and a negative impact on the likelihood of receiving CVC is expected. This is due to the investment objectives of both investors: Since VC is mainly motivated by monetary aspects, compensation is provided by cash flow streams, in contrast CVC also has a strategic motivation when investing in biotech companies; its compensation consists of both, cash flow streams and control rights.

Since the data has been collected from European and North American companies it is controlled for country specific effects on the likelihood of receiving VC and CVC. As Germany's and most of the other European countries' financial systems are bank dominated in comparison to the more stock market orientated UK or U.S. systems there are differences in CVC and VC activity expected. An active stock market is a predominant condition for a successful VC participation since going public is meant to be the most successful exit channel, with respect to monetary and non-monetary utility.[119] There is also empirical evidence for the frequency of VC and CVC in North America and Europe. The data provides evidence for VC and CVC being more likely in the United States than in the remaining participating European countries.[120] The control variables are operationalized as follows. The binary coded area variables equal one if the biotech company is located in the respective country: *area_US*: Company is located in North America; *area_UK*: Company is located in the United Kingdom; *area_GER*: Company is located in Germany. Furthermore it is controlled for the importance of receiving money of a strategic alliance partner. *Receiving_money* equals one if receiving money is very to extremely important for the company and zero otherwise. If payment terms are predominantly or completely preferred over control rights[121] when engaging in a strategic alliance the dummy *payment_terms* equals one, and zero otherwise. Table 3.1 presents an overview of the independent variables and their expected impact on the likelihood of receiving VC and CVC.

---

119 See Hochhold (2010).
120 See Dushnitsky (2006).
121 Control rights in this context refer for example to patent rights, marketing rights or distribution rights.

3.4.2.4 Descriptive Statistics

From all companies that have entered a strategic alliance, 65 companies have solely received VC, 38 exclusively CVC and 71 have received both, VC as well CVC. In general more companies have received VC than CVC, 41% are funded by both, VC as well as CVC.

The most important motivation for engaging in an alliance is the reputation of the cooperation partner, which 90% of companies in the sample considered as important. Reputation is followed by the importance of blocking competition by a cooperation partner, which is important for 72% of the companies. All other support variables are considered to be important by about half of the participating companies.

Table 3.2 shows the mean values of the independent variables subject to their respective capital provider. The mean comparison already indicates coherence in the demand for value-adding and VC / CVC funding. The experience in marketing and distribution of a cooperation partner is important for 54% of the companies that have received VC and for 29% companies that have not received VC. The partner's research experience is more often important for companies that have received CVC (59%) in comparison to companies that have not received CVC (45%), which would be in accordance with hypothesis 2 i). The partner's ability to block competition is more often a motivation for engaging in a cooperation for companies that are CVC financed (77%), which would already support hypothesis 3 b).

Table 3.1
Overview of Independet Variables and Expected Impact on Being VC or CVC financed

The following table shows the definition of the independent variables used in seemingly unrelated bivariate probit models as well as the expected impact on the likelihood of VC and CVC, whereas a '+' implies a positive and '-' a negative impact.

Variable	Variable Definition	VC	CVC
Support Variables			
Management experience	Dummy for management experience of partner is important	+	
Blocking competition	Dummy for role of partner in blocking competition is important	+	+
Reputation	Dummy for reputation of partner is important	+	+
Research experience	Dummy for research experience of partner is important		+
IP access	Dummy for access to intelectual property is important in partnership		+
Reducing costs	Dummy for reducing costs in partnership is important		+
Level of Information Asymmetry			
Early stage	Dummy for products being in a early development stage	+	+
Therapeutic	Dummy for therapeutic product development	+	-
Control Variables			
Receiving money	Dummy for receiving money from partner is important		
Payment terms	Dummy for prefering payment terms over control rights		
Area US	Dummy for company location in North America		
Area UK	Dummy for company location in the United Kingdom		
Area GER	Dummy for company location in Germany		

Table 3.1: Overview of Independent Variables and Expected Impact on Being VC or CVC financed.

The difference in the importance of research experience and the possibility of reducing costs when engaging with a strategic alliance partner are also significant. Both support variables are ranked as being important more often by portfolio companies that have not received VC than those companies which have received VC. For CVC financed companies the blocking of competition and the research experience of the alliance partner is more important than for non-CVC financed companies. This provides some descriptive evidence for hypotheses 2 and 3.

The mean age of the companies in the sample is 10 years. It can be seen that the average age of companies is smaller when funded by a VC in comparison to those funded by CVC. This is in line with market surveys that find that VC does invest earlier in companies than CVCs.[122] Further insights are, that 36% of the biotech companies are in an early development stage of their products. 75% of companies that are active in the therapeutic sector have received CVC whereas 59% of companies that have not received CVC are active in the therapeutic sector.

The descriptive analysis of the control variables show that receiving money from an alliance partner is important for companies more often that have received CVC than for companies that have not received CVC. If companies are willing to give up control rights in favour of additional payment terms VC becomes less likely. These findings concerning the descriptive statistic of the control variables point in the expected direction and provide evidence for the validity of the data.

---

122 See Dushnitsky (2006) and Gompers/ Lerner (1998).

Table 3.2
Descriptive Statistics

The following table presents the summary statistics for 174 biotechnology companies depending on VC and CVC investment. Asteriks indicate two-sample test of equl means as being significant at 1%***, 5%**, and 10%* levels.

N		Mean VC 136	Mean Non-VC 38	Mean CVC 109	Mean Non-CVC 65
Support Variables	Management experience	0.544	0.289 ***	0.450	0.554 *
	Blocking competition	0.721	0.737	0.771	0.646 **
	Reputation	0.919	0.868	0.890	0.938
	Research experience	0.500	0.658 **	0.587	0.446 **
	IP access	0.574	0.474	0.514	0.615 *
	Reducing costs	0.471	0.658 **	0.532	0.477
Level of Information	Early stage	0.375	0.316	0.367	0.354
Asymmetry	Therapeutic	0.699	0.658	0.752	0.585 ***
Control Variables	Receiving money	0.721	0.684	0.790	0.585 ***
	Payment terms	0.294	0.526 ***	0.349	0.338
	Area US	0.162	0.184	0.229	0.062 ***
	Area UK	0.147	0.263 *	0.184	0.154
	Area GER	0.360	0.395	0.339	0.415

Table 3.2: Descriptive Statistics

Table 3.3 illustrates the correlation matrix for the independent variables. All correlation coefficients are well below 0.4. Companies that rate blocking of competition as an important reason for cooperating also consider the reputation of the partner, research experience as well as IP access and reducing costs as important reasons, which is indicated by a positive correlation coefficient. The need for management experience is positively correlated with receiving money as a main motivation for entering a strategic alliance and is also weakly positive correlated with the preference for payment terms over control rights as well as with company location in the United States.

Table 3.3
Correlation of Independent Variables

This table shows the Cramer's V correlation coefficients of the binary coded independent variables for 174 biotechnology companies. Asteriks indicate variables as being significant at 1%***, 5%**, and 10%* levels.

		(1)	(2)	(3)	(4)	(5)
(1)	Management experience	1				
(2)	Blocking competition	0.1144	1			
(3)	Reputation	-0.0073	0.2041***	1		
(4)	Research experience	0.0131	0.1200*	0.0619	1	
(5)	IP access	0.0255	0.3745***	0.0331	0.2477***	1
(6)	Reducing costs	-0.0800	0.2714***	0.0869	0.2635***	0.3444***
(7)	Early stage	-0.0664	0.0637	0.0328	0.0318	0.0298
(8)	Therapeutic	-0.0651	0.1418**	0.0875	0.0713	0.0448
(9)	Receiving money	0.0362	-0.0510**	0.1935***	-0.0325	-0.1893***
(10)	Payment terms	-0.0075	-0.1203	-0.1039	0.0710	-0.0511
(11)	Area US	-0.0668	0.0690	0.0356	0.0155	0.1550**
(12)	Area UK	0.0409	-0.0927	-0.0127	-0.0011	-0.1393**
(13)	Area GER	-0.0540	-0.1159	-0.0872	-0.1722**	-0.1273*

		(6)	(7)	(8)	(9)	(10)
(7)	Early stage	-0.0293	1			
(8)	Therapeutic	0.0900	0.1952***	1		
(9)	Receiving money	-0.1124	0.0027	0.1505***	1	
(10)	Payment terms	0.0075	0.1579**	0.0685**	0.1935***	1
(11)	Area US	0.0051	-0.0160	0.2667***	0.1817***	0.0000
(12)	Area UK	-0.0714	0.0360	-0.0556	0.0881	0.0850
(13)	Area GER	0.0063	-0.0539	-0.1839***	-0.1214	-0.0268

Table 3.3: Correlations of Independent Variables

The correlation coefficient for being in an early stage of development and being active in the therapeutic sector is positive, whereas these two variables are negatively correlated to the age of the company. This finding is in line with the longer product development process in the therapeutic sector. In general 68% of the companies in the sample are active in therapeutics.

3.4.3 Methodology and Empirical Results

3.4.3.1 Seemingly Unrelated Bivariate Probit Model

The derived hypotheses are tested in a bivariate regression model.[123] A seemingly unrelated bivariate probit model allows controlling for any correlation among dependent variables and error terms, as the value of one variable will be dependent of the value of the other one. The dependent variables are binary coded as in whether or not the biotechnology company has received equity and from whom, VC or CVC. The probit model is nonlinear and the magnitude of the change in the outcome probability, for a given change in one of the independent variables, depends on the levels of all independent variables. Therefore the value of the estimated coefficients cannot be easily interpreted but the prefix indicates the direction of the relation between the dependent and independent variables.

The bivariate probit model is used to estimate a pair of probit models. The financing events are estimated simultaneously as it is assumed that the decision for VC and CVC are interrelated as companies may have received VC, CVC or both. Therefore axes of the two distributions will not be orthogonal.

---

123 For a detailed discussion of bivariate regression models see Long (1997); Long/ Freese (2006); Green (2000).

The general structure of the bivariate probit can be separated in two probit models with correlated error terms, which are estimated simultaneously:

$$I: y_{1i}^* = \beta_{1i}' x_{1i} + u_{1i},$$
$$y_{1i} = 1 \quad \text{if } y^*_{1i} > 0, \quad y_{1i} = 1 \text{ otherwise,}$$

$$II: y_{2i}^* = \beta_{2i}' x_{2i} + u_{2i},$$
$$y_{2i} = 1 \quad \text{if } y^*_{2i} > 0, \quad y_{2i} = 0 \text{ otherwise,}$$

$$\text{with} \quad \text{Cov}\left[\eta_i + \varepsilon_{1i}, \eta_i + \varepsilon_{2i}\right] = \rho$$

Y_1 and y_2 are the dependent variables VC and CVC financing, x_i are the independent variables and β_i the coefficient of the respective independent variable i. When the Covariance of the error terms is different from zero the error terms of each model are consistent of a part ε_i that is unique to that model and of a part η_i that is common to both models. In that case the dependent variables are interrelated and the simultaneous estimated bivariate model is superior to estimating two separate probit models.[124]

3.4.3.2 Empirical Results

Table 3.4 presents the results of three seemingly unrelated bivariate probit models. The main model includes all independent variables, model two estimates the impact of the support variables and model three the impact of the control variables on the likelihood of receiving VC respectively CVC. To address possible multicollinearity and endogeneity within the independent variables the data is

---

124 See Green (2000).

tested with the variance inflation factor (VIF). The VIF is a method of detecting the severity of multicollinearity issues. The inflation factor measures to what extent the variance of a coefficient is increasing due to collinearity. A common cut-off criteria for excluding a variable is VIF > 4. Whereby the square root of the variance inflation shows how much larger the standard error is, in comparison to what it would be if the variable was uncorrelated with the other independent variables in the equation.[125] The VIFs for all variables in the model are below 2, so that no variables have to be excluded for reasons of multicollinearity. All three models are highly significant. The test for superiority of the simultaneous estimation with the seemingly unrelated probit model is also highly significant and superior to estimating two independent probit models.

---

125 See Craney/ Surles (2002).

Table 3.4
Seemingly Unrelated Bivariate Probit Models

This table shows the estimates of the seemingly unrelated bivariate probit model of the impact of demand for advice and information asymmetry on the likelihood of reveiving VC / CVC. Standard errors are in parenthesis. The Wald χ^2-statistic tests the hypothesis whether all coefficients are jointly equal zero. When the correlation coefficient of the simultenous estimated probit models is significantly different from zero the bivariate model is superior to estimating separate probit models, which is indicated in p > χ^2 (II). Asteriks indicate variables as being significant at 1%***, 5%**, and 10%* levels.

	Model I				Model II				Model III			
	VC		CVC		VC		CVC		VC		CVC	
	Coef.	se	Coef.	se	Coef.	se	Coef.	se	Coef.	se	Coef.	se
Management experience	0.754	(0.246)***	-0.311	(0.215)	0.640	(0.238)***	-0.341	(0.204)*				
Blocking competition	-0.58	(0.359)*	0.911	(0.342)***	-0.323	(0.305)	0.826	(0.287)***				
Blocking competition later stag.	0.373	(0.340)	-0.277	(0.324)	0.306	(0.271)	-0.137	(0.245)				
Reputation	0.365	(0.405)	-0.961	(0.397)**	0.533	(0.384)	-0.740	(0.391)*				
Research experience	-0.693	(0.263)***	0.420	(0.246)*	-0.487	(0.246)**	0.434	(0.213)**				
IP access	0.595	(0.282)**	-0.677	(0.271)**	0.497	(0.255)*	-0.721	(0.246)***				
Reducing costs	-0.339	(0.266)	-0.188	(0.240)	-0.487	(0.253)*	0.120	(0.220)				
Early Stage	0.350	(0.312)	-0.158	(0.291)					0.171	(0.253)	-0.028	(0.218)
Therapeutic	0.126	(0.288)	0.226	(0.235)					-0.041	(0.258)	0.298	(0.227)
Receiving money	0.260	(0.272)	0.620	(0.248)**					0.249	(0.257)	0.460	(0.224)**
Payment terms	-0.471	(0.263)*	-0.139	(0.231)					-0.563	(0.233)**	-0.070	(0.219)
Area US	-0.460	(0.398)	1.043	(0.386)***					-0.425	(0.361)	0.881	(0.354)**
Area UK	-0.514	(0.382)	0.557	(0.322)*					-0.488	(0.354)	0.445	(0.307)
Area GER	-0.472	(0.357)	0.226	(0.267)					-0.328	(0.296)	0.144	(0.242)
Constant	0.828	(0.536)*	0.064	(0.473)	0.427	(0.397)	0.751	(0.397)*	1.090	(0.347)***	0.414	(0.276)
N	174				174				174			
Wald χ^2	109.74				30.42				32.22			
p > χ^2 (I)	0.000				0.007				0.004			
p > χ^2 (II)	0.000				0.000				0.000			

Table 3.4: Seemingly Unrelated Bivariate Probit Models

Venture Capitalist's support

The first VC model provides support for hypothesis 1, indicating a positive relation between management experience as an important motivation for cooperation and the likelihood of receiving VC financing. The reliability of the model is validated in the second model as the impact is still highly significant.

There is no evidence for hypotheses 3 a) and 4 a) assuming a positive relation between the likelihood of receiving VC and the importance of support in blocking competition and engaging with a reputational partner.

An additional finding is the importance of access to intellectual property that leads to an increase in the likelihood of receiving VC. This might result from the institutional structure of VC funds.[126] As VCs monitor and support companies they have access to detailed information about a companies strategy and progress. Portfolio companies in a VC fund are often pooled in the same industry with similar research focus. The value of the portfolio can be enhanced by funding companies that fit well in a strategic sense. The VC's ability to find complementary resources in its portfolio to arrange collaboration between the portfolio company's synergies can be used to optimize portfolio returns. At the same time VC is able to punish moral hazard and hold up problems within its portfolio companies by exchanging research know-how and providing access to intellectual property.[127] In doing so, knowledge transfer and

---

126 Portfolio companies financed by VC are pooled to VC funds. The fund is monitored and supported by a team of fund managers that have unique knowledge about strategy and progress of their financed companies. For more details on the institutional structure on VC funds see Gompers/Lerner (2006).
127 See Hellmann (2002) and Lindsey (2008).

acess to intellectual property can be stimulated inside a VC fund's companies.

Corporate Venture Capitalist's Support

In line with hypothesis 2 is the increasing likelihood for CVC with a demand for accessing tangible and intangible research related resources. This relation is significantly supported by the positive impact of the demand for an experienced partner in conducting research on the likelihood of receiving CVC in model I and II. There is no evidence for the expected positive impact of the need for reducing costs on the likelihood of receiving CVC. The likelihood for CVC decreases with the need for access to intellectual property of the portfolio company, which is in contrast to the previous theoretical assumption. It might be the case that access to intellectual property provided by pharmaceutical companies is generally provided by strategic alliances.[128] If that is the case CVC has no additional incentive providing external additional strategic resources that would be triggered by the equity stake. In contrast VC as a pure financier has the incentive to provide support in finding a suitable strategic alliance partner in optimizing its fund performance.

If blocking competition is an important motivation for engaging in an alliance the likelihood of receiving CVC is increasing. This finding is in line with hypothesis 3 b) and robust for model I and II.

In contrast to hypothesis 4 b) the likelihood for CVC financing is decreasing with demand for engaging with a reputational partner. This also might be due to a prior strategic alliance engagement and satisfaction certification transfer.[129]

---

128 See Wadhwa/ Phelps (2010).
129 See Gulati/ Higgins (2003).

Level of Information Asymmetry

The age of the portfolio company has a significant negative impact on the likelihood of receiving VC and a positive impact on the likelihood of receiving CVC. This means that companies receive VC at an earlier stage, which is also in line with a previous empirical study of Gompers and Lerner (2006). The negative relation between VC and the portfolio company's age is in line with hypothesis 5 a). Hence, VC is more likely with a high level of information asymmetry. There is no further support for hypothesis 5. Both the stage of product development and the product development focus on therapeutics have no statistical impact on the likelihood of receiving VC or CVC.

Control Variables

The likelihood for CVC financing is increasing if receiving money is an important motivator for engaging in a strategic alliance. This intuitive finding indicates that the motivation for engaging in a strategic alliance can be used as an appropriate measure estimating the demand for support.

There is a positive impact on the likelihood of receiving VC if the portfolio company prefers control rights over payment terms. This is in accordance with the institutional definition of VC and CVC. Since VC is having a solely monetary objective CVC follows also strategic objectives.[130] Consequently, a portfolio company should prefer VC over CVC if there is a bias on control rights.

The likelihood of receiving CVC and VC is dependent on the portfolio company's location. CVC and VC are more common for companies located in North America.[131] This is in line with the posi-

---

130 See Tykvova (2003) and Dushnitsky (2006).
131 See Dushnitsky (2006).

tive impact of a North American company location on the likelihood of receiving CVC.

3.4.4 Discussion of Empirical Results

The empirical multivariate results of the seemingly unrelated probit models are robust concerning the used statistical model and the model fit. Nevertheless, the study is subject to some limitations with respect to the interpretation and generalization of the results. This is due to the character of the data used.

The sample is restricted to data of biotech portfolio companies since strategic alliances and CVC are most common in the pharmaceutical industry. The advantage of an industry focus is that cross-industry effects can be neglected, which has been criticized in related literature.[132] The downside of focusing the study on a particular industry is that empirical findings cannot be generalized for other sectors. The speciality of the biotech industry is long development cycles combined with high uncertainty about product success. Other industries, for example computers or telecommunications do not have the same characteristics specified before.

A limitation of the study is the possibility of receiving CV or CVC after the online survey took place. This limitation is more severe for companies that have not yet received CVC, since companies that have already received CVC are on average younger than companies that have not. This is not the case for VC financed companies who are, on average, younger than companies that have not received VC. Another limitation is the predominance of German biotech companies since this does not represent the actual dispersion of biotech companies in the sample regions. Therefore the re-

---

132 See Dushnitsky (2006).

sults could be distorted as support activity varies across countries.[133]

The independent variables measuring support demand used are determined by a proxy for the motivation for entering a strategic alliance. The reliability of the proxy has been positively tested by the control variables *receiving money* and *payment term*s. Nonetheless, the demand is not measured directly and therefore additional information might not be reproduced.

Resulting from the cross-sectional sample it cannot be controlled for unobserved heterogeneity and time precedence.

3.5 Conclusion

The study provides evidence for the efficient allocation of capital from a portfolio company's perspective. Consequently, a company's specific demand for support in certain areas does have a significant impact on the likelihood of receiving VC and CVC.

These findings contribute to the literature on CVC's and VC's value-adding activity. Previous studies have analysed the value-adding contribution of VC and CVC on firm success and performance from an ex post perspective.[134] This study sheds light on the impact of an ex ante demand for value-adding on the likelihood of receiving VC and CVC. According to the already existing literature on value adding-activity it is assumed that VC provides more effective support in management related areas whereas CVC support is more efficient in research related areas. According to a model of Casamatta (2003) the equity stake of a capital provider ensures that

---

133 See Hochhold (2010).
134 See Gorman/ Sahlman (1989); Maula et al. (2005) and Hellmann (2002).

value-adding is efficiently provided by the institution whose effort is most efficient.

The paper shows that VC financing is more likely if the portfolio company is considering management experience of a cooperation partner as being important. In contrast, CVC financing is more likely if research experience is an important motivator for engaging in a strategic alliance. There is actual evidence that the importance of research related support decreases the likelihood of VC financing. Given this, the determining factor representing the effective capital allocation is the demand for support in certain areas, regardless of the supply and demand for VC and CVC.

Contrary to the previously derived assumptions, there is no evidence for the positive impact of the demand for a reputed partner on the likelihood of receiving VC and CVC. Furthermore, it was not possible to show the likelihood of VC is increasing with a demand for blocking competition. There was no evidence for a positive relation between a demand for reducing costs and IP-access to the likelihood of receiving CVC. The missing or contrary results might evolve from the previous strategic alliance engagement of the studied biotech companies. It could be the case that some support is already provided by an external company without an equity stake.

For future research, it would be interesting to investigate the interdependency of the value adding ability of strategic alliances, VC and CVC in more detail. Further research could, for example, focus on the time structure of engaging with each partner, the detailed relation between the involved parties and the time and mode of the exit.

Bibliography for Chapter 3

Aghion, Philippe / Tirole, Jean (1994): The Management of Innovation, in: Quarterly Journal of Economics, Vol. 109, No. 4, 361-379.

Al-Laham, Andreas / Schweizer, Lars / Amburgey, Terry (2010): Dating Before Marriage? Analyzing the Influence of Pre-Acquisition Experience and Target Familiarity on Acquisition Success in the "M&A as R&D" Type of Acquisition, in: Scandinavian Journal of Management, Vol. 26, 25-37.

Allen, Jeffrey / Philipps, Gordon (2000): Corporate Equity Ownership, Strategic Alliances, and Product Market Relationships, in: Journal of Finance 55, 2791-2815.

Anand, Bharat / Galetovic, Alexander (2000): Weak Property Rights and Holdup in R&D, in: Journal of Economics and Management Strategy, Vol. 9, No 4, 615-642.

Casamatta, Catherine (2003): Financing and Advising: Optimal Financial Contracts with Venture Capitalists, in: Journal of Finance 58, 2059-2086.

Chung, Seungwha / Singh, Harbir / Lee, Kyungmook (2000): Complementary, Status Similarity and Social Capital as Drivers of Alliance Formation, in: Strategic Management Journal 21, 1-22.

Cestone, Giacinta / White, Lucy (2003): Anticompetitive Financial Contracting: The Design of Financial Claims, in: Journal of Finance 58, 2109-2141.

Craney, Trevor / Surles, James (2002): Model-Dependent Variance Inflation Factor Cutoff Values, in: Quality Engineering 14, 391-403.

Dushnitsky, Gary (2006): Corporate Venture Capital: Past Evidence and Future Directions, in: Mark Casson et al. (ed.), The Oxford Handbook of Entrepreneurship, Oxford University Press, 387-431

Dushnitsky, Gary / Shapira, Zur (2009): Entrepreneurial Finance Meets Organizational Reality: Comparing Investment Practices by Corporate and Independent Venture Capitalists, Working Paper, University of Pennsylvania, August 2009.

European Commission (2009): The Financing of Biopharmaceutical Product Development in Europe, Final Report, Copenhagen / Brussels, October 2009.

Gompers, Paul (1996): Grandstanding in the Venture Capital Industry, in: Journal of Financial Economics 42, 133-156.

Gompers, Paul / Lerner, Josh (2000): The Determinants of Corporate Venture Capital Success: Organizational Structure, Incentives, and Complementaries, in: Randall Morck (ed.), Concentrated Ownership, University of Chicago Press, 17-54.

Gompers, Paul / Lerner, Josh (2001): The Venture Capital Revolution, in: Journal of Economic Perspective 15, 145-168.

Gompers, Paul / Lerner, Josh (2006): The Venture Capital Cycle, 2nd Edition, MIT Press.

Gorman, Michael / Sahlman, William (1989): What Do Venture Capitalists Do?, in: Journal of Business Venturing 4, 231-248.

Green, William (1993): Econometric Analysis, 2nd Edition, Macmillan Publishing Company.

Gulati, Ranjay / Higgins, Monica (2003): Which Ties Matter When? The Contingent Effects on Interorganizational Partnerships on IPO success, in: Strategic Management Journal 24, 127-144.

Hellmann, Thomas (2002): A Theory of Strategic Venture Investing, in: Journal of Financial Economic 64, 285-314.

Hellmann, Thomas / Puri, Manju (2002): Venture Capital and the Professionalization of Start-up Firms: Empirical Evidence, in: Journal of Finance 57, 663-691.

Higgins, Matthew / Rodriguez, Daniel (2006): The Outsourcing of R&D through Acquisitions in the Pharmaceutical Industry, in: Journal of Financial Economics, Vol. 80, 351-383.

Hochhold, Stefanie (2010): Monitoring and Support in Venture Capital Financing, Working Paper, University of Munich, July 2010.

Ivanov, Vladimir / Xie, Vei (2010): Do Corporate Venture Capitalists Add Value to Start-up Firms? Evidence from IPOs and Acquisitions of VC-Backed Companies, in: Financial Management 39, 129-152.

La Porta, Rafael / Lopez-De-Silanes, Florencio / Shleifer, Andrei / Vishny, Robert (1998): Law and Finance, in: Journal of Political Economy 106, 1113-1155.

Lindsey, Laura (2008): Blurring Firm Boundaries: The Role of Venture Capital in Strategic Alliances, in: Journal of Finance 63, 1137-1168.

Lerner, Josh / Merges, Robert (1998): The Control of Technology Alliances: An Empirical Analysis of the Biotechnology Industry, in: The Journal of Industrial Economics 46, 125–156.

Long, Scott (1997): Regression Models for Categorical and Limited Dependent Variables, in: Advanced Quantitative Techniques in the Social Sciences, Vol. 7, Thousand Oaks.

Long, Scott / Freese, Jeremy (2006): Regression Models for Categorical Dependent Variables Using Stata, 2nd Edition, Texas, Stata Press.

Maula, Markku (2001): Corporate Venture Capital and the Value-added for Technology-based New Firms, Doctoral Dissertation Helsinki University of Technology.

Maula, Markku / Autio, Erkko / Murray, Gordon (2005): Corporate Venture Capitalists and Independent Venture Capitalists: What do They Know, Who do They Know and Should Entrepreneurs Care?, in: Venture Capital 7, 3-21.

Megginson, William / Weiss, Kathleen (1991): Venture Capitalist Certification in Initial Public Offerings, in: Journal of Finance 56, 879-903.

Orman, Cuneyt (2009): Corporate R&D, Venture Capital, and Capital Markets, SSRN Working Paper, 2009.

Rosenstein, Joseph / Bruno, Albert / Bygrave, William / Taylor, Natalie (1993): The CEO, Venture Capitalists, and the Board, in: Journal of Business Venturing 8, 99–113.

Sapienza, Harry (1992): When Do Venture Capitalists Add Value, in: Journal of Business Venturing 7, 9-27.

Sapienza, Harry / Manigart, Sophie / Vermeir, Wim (1996): Venture Capital Governance and Value Added in Four Countries, in: Journal of Business Venturing 11, 439-469.

Sykes, Hollister (1990): Corporate Venture Capital: Strategies for Success, in: Journal of Business Venturing 5, 37-47.

Stuart, Toby / Hoang, Ha / Hybels, Ralph (1999): Interorganizational Endorsements and the Performance of Entrepreneurial Ventures, in: Adminsitrative Science Quarterly 44, 315-349.

Teece, David (1986): Profiting From Technological Innovation: Implications for Integration, Collaboration, Licensing and Public Policy, in: Research Policy 15, 285-305.

Tykvová, Tereza (2007): What Do Economists Tell Us About Venture Capital contracts?, in: Journal of Economic Surveys 21, 65-89.

Wadhwa, Anu / Phelps, Corey (2010): An Option to Ally: A Dyadic Analysis of Corporate Venture Capital Relationships, SSRN Working Paper, February 2010.

Zider, Bob (1998): How Venture Capital Works, in: Harvard Business Review, November – December 1998, 131-139.

4 Corporate Venture Capital Performance – Is There a First-Mover Advantage?[135]

4.1 Introduction

For established high-tech companies innovation is an eminent condition for growth. Corporations use various approaches to innovate, including internal R&D, incubation of new businesses, and strategic investments and alliances. Affected by growing venture capital (VC) activity in the last few decades, corporations have institutionalized corporate venture capital (CVC) as an alternative approach to innovation.

CVC is defined as a minority equity investment by established, often large firms, in smaller, research-intensive entrepreneurial companies ("portfolio company").

A CVC investment is determined by its objective and the degree to which the operations of the investing company and the portfolio company are linked. Two often cited objectives or motivations for investing via CVC are strategic and financial.[136] Strategic investments seek to exploit synergies with the portfolio company, while financially motivated investments seek attractive financial returns.[137]

---

135 This chapter is joint work with Carolin Haeussler, Munich School of Management, University of Munich, and Matthew Higgins, College of Management, Georgia Institute of Technology, Atlanta.
136 See for example, Corporate Strategy Board (2000); Asset Alternatives (2002); PriceWaterhouseCoopers (2006); MacMillan et al. (2008).
137 See Chesbrough (2002).

The theoretical definition of financial return is straight forward since only a monetary dimension, determined by the variance of the portfolio company's value over time, is considered. In contrast, the definition of strategic or technological return is more complex. In related literature, technological return consists of various types of benefits for the investing company. For example, technological benefits for the investing company may include: The provision or spill over of new ideas from the portfolio company;[138] gaining a window of new or future technology; the ability to execute real options on new technologies; or the formation of a strategic alliance with or acquisition of the portfolio company[139].

Interestingly, while the use of CVC investments by firms has increased in the past few decades, little empirical evidence exists on realized financial or technological returns to CVC investors. A few notable exceptions that examine financial returns include Allen and Hevert (2007) who find that for the U.S. information technology sector financial returns exceed the investing company's cost of capital in 39% of the cases. Gompers and Lerner (2006) find that CVC investments that have a high strategic fit with their portfolio companies are more successful than comparable VC investments in terms of the likelihood of taking the portfolio company public. Stuart et al. (1999) find that CVC financed companies go public faster and with higher valuations than non CVC-backed companies. Finally, Maula and Murray (2002) as well as Ivanov and Xie (2010) show that CVC and VC co-financed companies have higher market valuations than purely VC financed companies. An overarching theme in these studies is that the presence of a CVC investor adds value in some form or another to the portfolio firm that is eventually captured in financial returns.

---

138 See Dushnitsky/ Lenox (2006) and Wadhwa/ Kotha (2006).
139 See Li et al. (2009); Gompers/ Lerner (2006).

Complementing these studies are a few focused on technological returns. Wadhwa and Phelps (2010) find that technological return is negatively related to a portfolio company's stage of development at the time of investment and positively related to the age of the portfolio company. In contrast, Smith (2009) demonstrates that technological returns in the medical device industry are greater in earlier stages of development and, in cases when the entrepreneur is a clinician, the profession of the entrepreneur has a positive impact on technological or innovation returns. Dushnitsky and Lenox (2006) find that weak intellectual property regimes and a CVC investor's absorptive capacity have a positive impact on technological return. Finally, Wadhwa and Kotha (2006) find that technological return is non-linearly dependent upon the number of CVCs investing in a particular portfolio company and the CVC's involvement.

The initial evidence provided by these empirical studies seems to suggest that, in general, CVC investments have the ability to create financial as well as technological value for the investing company. Additionally, there is evidence to suggest that returns to CVC investors may be superior to VC investors.[140] Research has demonstrated that, on average, early stage VC investments yield lower returns than later stage investments.[141] This is in contrast to corporate finance theory, which argues that risk-averse investors should receive a premium for early stage investments due to higher investment risk. We analyse whether CVC returns are more in line with finance theory or the extant empirical VC literature. Furthermore, in most industries, CVC investments are occurring in a multi-partner investment setting. That is, several CVC investors may be investing sequentially or simultaneously, at potentially dif-

---

140 See Maula/ Murray (2002) and Ivanov/ Xie (2010).
141 See Cumming (2008) and Cumming/ Walz (2004).

ferent stages, in a particular portfolio company. Given this setting the question we address is whether the timing of entry of CVC investment has an impact on an investor company's (financial or technological) performance.

This consideration is drawn upon three approaches determining the impact of timing of investment on CVC performance. Firstly, we expect a risk-adjustment to CVC returns, meaning that risk-averse investors should receive a premium as compensation for increased risk emanating from early-stage CVC investments. Secondly, we expect to see a first investor advantage in technological returns when determining the strategic path development of a portfolio company. And third, we expect a premium for investors that send an initial positive signal on the portfolio company's quality to the capital market, which will be reflected in the financial return.

Based on a dataset consisting of 689 CVC financing rounds in over 300 biotechnology companies we estimate the impact of CVC's entry in a multi-partner investment setting on overall CVC investment return, technological return and financial return. Our dataset is unique in that it includes detailed information about financial and liquidity events in the life span of our companies as well as information to proxy the quality of the companies. We find support for our notion of a first-mover advantage. Being the first CVC investor is significantly positively related to both technological and financial return. In addition, investing early in a biotechnology company pays off in terms of financial return but not when seeking technological return.

Using a unique data set, this study makes three main contributions to the literature. Firstly, we extend the empirical CVC literature. As mentioned above, a limited number of studies explore one particular dimension of CVC investment returns either technological or financial. Our setting, however, allows us to study both dimensions of CVC performance simultaneously. We find that, on average, CVC returns indeed consist of technological as well as on financial returns. Secondly, we contribute to the literature identifying determinants of overall CVC returns. Finally, we contribute to the broad market entry literature. More specifically, we find a relation between the timing of a CVC investment in a multi-partner setting and investor returns and, in particular, we identify a first-mover advantage.[142]

4.2 Corporate Venture Capital Returns

4.2.1 Determinants of Corporate Venture Capital Returns

The determinants of CVC performance are not well documented in theoretical literature. However we attempt to draw parallels from the very comprehensive VC literature in hopes of identifying factors that may help determine or predict CVC investment performance. We acknowledge that given the potential different underlying strategic motivations for making CVC investments, the corollary with VC may only exist when motivations are consistent.

CVC and VC investors fill a similar role in financing and advising entrepreneurial companies;[143] however, CVC funds differ in their organizational and incentive structure. For example, CVC funds are often structured as a corporate subsidiary and have much lower incentive-based compensation. According to theory the structure of VC funds, in particular the reliance on limited partnerships of finite life with substantial profit sharing, is critical to suc-

---

142 Lavie et al. (2007) discuss a first-mover advantage within the context of strategic alliance.
143 See Hellmann (2002).

cess. For example, VC funds have incentives to terminate underperforming companies in their portfolio, due to the VC fund's finite life, or supply sufficient effort in nurturing their portfolio companies, since the fund's management is participating in resulting returns.[144] The inferior incentive-scheme for CVC funds, due to a less efficient CVC fund structure may be offset by benefits emanating from superior information and complementarily activities with a portfolio company. For example, pharmaceutical companies, in their role as CVC investors, have a comparative advantage in information production due to their superior scientific and commercial knowledge relative to VCs and other market investors. When investing in nascent, research-intensive companies CVC investors may be able to select better firms and may add greater value to their portfolio companies by using information from their related lines of business.[145]

Finally, since the motivation to make a CVC investment may differ from the purely financial VC objective, a CVC investor may have stronger incentives to invest in more information gathering than a VC who takes only a partial, and temporary, equity stake. This can lead to a comparative advantage in selection for CVC investors over VC investors and other market investors. This superior selection ability may compensate for their less efficient organizational structure.

Previous research has pointed out that VC performance is dependent on two main areas: Micro-factors, such as the structure of limited partners, the VC firm or the portfolio company and macro-

---

[144] Please refer for a brief discussion of the importance of VC structure on fund performance to Gompers/ Lerner (2006).
[145] See Gompers/ Lerner (2006) and Danzon et al. (2005).

factors, including environmental determinants.¹⁴⁶ Therefore the timing of a VC investment is of special interest when analysing VC performance. According to corporate finance theory there is a positive relationship between the risk of an investment and the return required by an investor.¹⁴⁷ A product's stage of development is a common and important proxy used to estimate various risk dimensions in the early stage financing literature. Consequently, at the company-level, VCs will require a higher return for an earlier stage investment than if they invested at a later stage.¹⁴⁸ This is in contrast to findings in previous empirical studies, which have shown that, on average, early stage VC investments yield lower returns than later stage investments¹⁴⁹ and that early stage VC investment have a negative impact on the likelihood of a successful exit.¹⁵⁰ As such, we are left with an inconsistency between theory and empirics with respect to VC investment performance.

Given the aforementioned discussion it is fair to suggest VC and CVC returns cannot be easily compared given CVC returns tend to be a function of two different determinants. While there are parallels between these two types of investment, the question arises whether CVC returns are more in line with theoretically predicted risk-adjusted returns than VC investments. According to Gompers and Lerner (2006) CVC investors have a comparative advantage over VC investors due to their superior selecting ability. This translates into a positive, value-adding effect when investing in complementary technological research areas. As a result, this compara-

---

146 For a literature overview on determinants of VC performance please refer to Söderblom/ Wiklund (2006).
147 See Brealey/ Myers (1996).
148 See Manigart et al. (2002).
149 See Cumming (2008) and Cumming/ Walz (2004).
150 See Hege et al. (2003); Das et al. (2003).

tive advantage might also produce a different distribution of CVC performance by time of investment than has already been observed for VC returns. Superior returns for early investors might be suggestive of the existence of some type of first-mover advantage for these investors; an issue which we discuss next.

4.2.2 Timing of Corporate Venture Capital Returns

4.2.2.1 Implications of Corporate Finance Theory

Corporate finance theory implies the timing of a CVC investment is important insofar as CVC returns should be risk-adjusted. In other words, risk-averse investors should prefer safe investments over identical riskier ones. As such, due to an investor's risk aversion they expect a return that appropriately compensates them for the project risk of a given investment. Risk-adjusted returns should therefore consist of a risk-free interest rate plus an appropriate risk premium, which is positive and dependent upon the investment's risk.[151] Consistent with the extant VC literature, a portfolio company's age and stage of product development are commonly used proxies for risk; younger firms and those with earlier stage technologies are viewed with more risk. To compensate for this increased risk we should see similarly higher CVC rates of return for early stage compared to later stage investments. Consequently, when accounting for overall CVC returns we expect a higher rate of return for early CVC investors since they face higher levels of uncertainty and should therefore demand a larger risk premium.

---

151 See Brealey/ Myers (1996).

4.2.2.2 Implications of Value-adding

Theorists have long demonstrated competitive advantages accruing to first movers[152] and empirical studies across industries have corroborated many of these advantages. Traditionally, first-mover advantages have been studied in product markets with extensions into organizational innovations[153] and managerial innovations.[154] Within our context a first-mover advantage derives from two potential sources. First, CVC investors may be able to gain an advantage over their rivals through the pre-emption of scarce resources and technology.[155] If a CVC has superior information they may be able to purchase an equity stake in promising technology at prices below those that may later prevail in the evolution of the market. Second, being a first-mover gives the CVC the opportunity to nurture the company and influence the strategic direction of the company. For example, studies have shown that it is mostly early network partners who set the path to an effective organization and company success.[156] In a multi-partner alliance setting, Lieberman and Montgomery (1988) have shown that a first-mover derives advantages from path dependence, efficient governance and longer lead time. Since CVC investors nurture and cultivate their portfolio companies during their investment period, being the first CVC investor provides them with the maximum opportunity to influence a portfolio company's development and potentially direct them in a particular strategic direction, which may be more beneficial for the

---

152 For example see Alpert et al. (1992); Carpenter/ Nakamoto (1989); Gal-Or (1985); Gilbert/ Newberry (1985); Golder/ Tellis (1993); Haines et al. (1989); Kardes/ Kalyanaram (1992); Kardes et al. (1993); Lieberman/ Montgomery (1988); Schmalansee (1982).
153 See Teece (1980).
154 See Chandler (1977).
155 See Lieberman/ Montgomery (1998).
156 See for example Hannan et al. (1996).

CVC investor. This view is shared by Wernerfelt and Karnani (1987) who argue that early entry or in this case, first-entry, is more beneficial to firms that can influence the way uncertainty surrounding the portfolio company is resolved. Therefore, we propose that a first-mover advantage exists in terms of technological return.

In contrast, CVC investments made at a later stage would not be conferred with these same opportunities. CVC investors would have a more difficult time changing the strategic path of a company. Moreover, later investors would be faced with a lower risk profile than earlier investors. That said, later stage investors might be able to free-ride on the first-mover's investment.[157] While it may be the case that later investors may experience some form of indirect information spill overs in R&D[158], free-riding is more likely to occur with the financial returns. If the first-mover is successful in their nurturing of a portfolio company one would expect fast-followers to be able to engage in some type of free-riding, especially in terms of financial returns.

4.2.2.3 Implications of Corporate Venture Capital Investment Signalling

In imperfect capital markets valuing entrepreneurial companies that lack a previous track record is difficult. In this case investors often rely on signals such as founder background[159], patents[160] or prominent partners[161] in order to proxy for quality. If CVC investors have superior capabilities which allow them to better evaluate

---

157 See Lieberman/ Montgomery (1988).
158 See Spence (1984); Baldwin/ Childs (1969).
159 See Eisenhardt/ Schoonhoven (1990), Burton et al. (2002) and Shane/ Stuart (2002).
160 See Hsu/ Ziedonis (2006) and Haeussler et al. (2009).
161 See Stuart et al. (1999), Baum/ Silverman (2004).

portfolio companies compared with the financial markets[162] then portfolio companies that have already received CVC investment should receive higher valuations relative to comparable companies not yet financed by a CVC.

Companies that benefit from a higher valuation obtained via a validation signal due to CVC investment are willing to pay for this signal. As a result, these companies should be willing to accept a discount for their initial CVC investment since their market valuation in the next financing rounds will subsequently, due to reduced information asymmetries, increase. This leads to a discount in the company's market valuation for first investing CVC companies. For example, Danzon et al. (2005) find that portfolio companies entering their first equity alliance[163] receive an average discount of 47% in valuation compared to companies that signed at least two prior alliance deals. The discount in valuation was still 28% when companies entered their second equity alliance. If an equity alliance provides a comparable signal to financial markets about CVC investments the following assumption can be drawn: If there is a discount for first CVC deals, then these investments should have a positive impact on financial returns.

4.3 Data Set

4.3.1 Sample Selection

To investigate the returns to CVC investors we assemble data from a variety of public and private sources. The primary data for this study comes from Deloitte Recap (Recap), a U.S. based biotechnology consulting company. Recap's proprietary data consists of a full

---

162 See Danzon et al. (2005) and Gompers/ Lerner (2006).
163 In the course of this study an equity alliance does not match with CVC entirely as CVC can also be provided without an alliance agreement.

set of round-by-round financings of over 600 biotechnology companies beginning in the early 1980s. The data includes valuation histories, investors and liquidation events. Relative to other sources of pre-money values such as VentureOne and VentureXpert, one of Recap's distinctive features is that for each financing round it reports the price per share, the number of pre-money shares outstanding and the number of shares issued.[164] This dataset has been complemented with Recap alliance data, containing alliance agreements including equity investments, acquisitions and licensing deals. In addition, for the biotechnology companies listed in Recap we compiled data on USPTO issued patents from the National Bureau of Economic Research (NBER) patent database. Finally, we gathered data on monthly mergers and initial public offerings (IPOs) from Securities Data Company (SDC).

The unit of analysis of our dataset is the CVC investor – company dyad. These dyads include all CVC investor – portfolio (biotechnology) company relationships which were entered prior to a first liquidation event (IPO, acquisition, or bankruptcy) of the portfolio biotechnology companies. While it is possible for CVC investors to invest in multiple rounds of the same portfolio company, this is not a common occurrence. Of the 689 CVC investor-company dyads, 577 are unique CVC investor-company pairs.

4.3.2 Variables and Descriptives

Table 4.1 provides a tabular overview of the variables along with their definition. Table 4.2 provides summary statistics for all of our variables.

---

[164] A portion of this data was also utilized by Hand (2007) in his study of determinants of round-by-round returns to pre-IPO VC investments.

Table 4.1
Overview of Variables

The following table shows the definition of the dependent and independent variables used in the empirical analysis.

	Variable	Variable Definition
Dependent Variables	Technological return	Dummy for (future) alliance with license agreement with CVC or acquisition of portfolio company by CVC
	Financial return	Annualized return on investment
	Return index	Index for cumulated technological and financial return
Independent Variables	CVC round A	Dummy for CVC investmet in financing round A
	CVC round B	Dummy for CVC investmet in financing round B
	CVC round C	Dummy for CVC investmet in financing round C
	CVC round D-K	Dummy for CVC investmet in financing round D or later
	First CVC investment	Dummy for first CVC investment in a particular company
	Company age at CVC investment	Age of portfolio company at time of CVC investment in month
Control Variables	Liquidation event=acquisition	Dummy for first liquidation event of portfolio company is a acquisition
	CVC stock	Number of CVC investments prior to a particular investment in portfolio company
	Equity stake	Equity stake CVC owns at time of investment
	Valuation round A	Valuation of portfolio company at time of financing round A in million $
	Alliance stock	Number of portfolio company's alliances previous to CVC investment
	Patent stock	Number of portfolio company's patents previous to CVC investment
	VC investment	Dummy for portfolio company has received VC
	VC co-investment	Dummy for CVC-VC syndication
	IPO activity	Number of overall U.S. IPOs in the month of CVC
	M&A activity	Number of overall U.S. acquisitions in the month of CVC disinvestment

Table 4.1: Overview of Variables

4.3.2.1 Dependent Variables

Technological Return

Within the empirical literature the operationalization of technological returns from CVC investments varies. For example, Wadhwa and Phelps (2010) define technological return as entering a strategic alliance with a previous CVC investor. Smith (2009) uses citations by the CVC investor to the portfolio company's patents, reflecting the relevance of the intellectual property to the CVC investor as a proxy for technological return. Wadhwa and Kotha (2006) as well as Dushnitsky and Lenox (2005) define technological return as "knowledge creation" through successful patent applications of the CVC investor. Finally, in a study focused on the relationship between financial and technological returns, Kang and Nanda (2010) define technological returns as the residual from a productivity regression.

We define the variable *technological return* as a dummy that equals one if a CVC investor either entered into a licensing agreement with its portfolio company or acquired it after the CVC investment. This definition partly corresponds with Wadhwa and Phelps (2010). They define "technological return" as an alliance with a CVC investor at any time after the CVC invested in the portfolio company. While they consider any alliance[165] as technological return, we restrict our technological return variable to licensing agreements and acquisitions in which indeed technology is transferred to the CVC investor. As we report in Table 4.2, 20% of the CVC investors in our

---

[165] Since alliances have various resource-related contents not all alliances can be considered of gaining a technological return. We restrict our analysis to alliances in which technology is acquired by the incumbent firm.

sample experience a technological return from their investment.[166] This number is slightly higher, but generally consistent, with the 16% reported in the HBM Partners Biotech M&A survey.[167]

Financial Return

We follow Kang and Nanda (2010) and estimate the variable *financial return* by the annualized return on CVC investment (ROI). The annual return on investment is calculated as:

$$ROI_i = \frac{P_{i,1} - P_{i,0}}{P_{i,0} \times Y_i}$$

where ROI is the annualized return of investment i, P_0 is the purchasing price per share at time of investment, P_1 is the price per share at time of disinvestment and Y is the investment period in years. The price per share at the time of disinvestment is based on the valuation of the first liquidation event: Bankruptcy, acquisition or IPO.[168] The mean (median) return on investment is close to 200% (46%) per year, which corresponds to a triplication of the in-

---

166 A number of previous studies have used patent information (for exaple Dushnitsky/ Lenox (2005); Schildt et al. (2003); Wadwha/ Kotha (2006) or compounds/drugs within a company's research pipeline to measure technological success. However, a statistical relationship may simply be spurious due to the presence of latent variables which are causal for CVC investment and patenting. More importantly, it is challenging to single out the effects of CVC investments on these measures.
167 See HBM Pharma (2009).
168 In the case of an IPO we use the IPO offering share price and in the case of acquisitions we calculate share price based on the acquisition price divided by total company shares. In the case of bankruptcy, share price is set at zero.

vestment per annum. The standard deviation indicates that the distribution of returns is widely dispersed.

Table 4.2
Summary Statistics

The table presents the summary statistics for 689 CVC investments. (d) indicates a binary variable.

	Variable	Mean	Std. Dev.	Min	Max
Dependent Variables	Technological return (d)	0.20	0.40	0	1
	Financial return	2.00	14.31	-0.73	337
	Return index	1.70	0.62	1	3
Independent Variables	CVC round A (d)	0.20	0.40	0	1
	CVC round B (d)	0.21	0.41	0	1
	CVC round C (d)	0.20	0.40	0	1
	CVC round D-K (d)	0.16	0.37	0	1
	First CVC investment (d)	0.55	0.50	0	1
	Company age at CVC investment	39.36	28.49	2	171
Control Variables	Liquidation event=acquisition (d)	0.10	0.31	0	1
	CVC stock	4.38	6.58	1	44
	Equity share	0.17	0.16	0.01	1
	Company valuation round A	8.16	11.02	0.01	105.39
	Alliance stock	3.70	5.52	0	76
	Patent stock	3.18	7.62	0	74
	VC investment (d)	0.67	0.47	0	1
	VC co-investment (d)	0.17	0.38	0	1
	IPO activity	44.04	22.81	3	101
	M&A activity	830.41	257.22	121	1421

Table 4.2: Summary Statistics

Return Index

In an effort to capture both returns simultaneously we generate the composite variable *return index*. The variable is ordinal scaled and equals one if neither a technological, nor a financial return has been obtained, equals two if either a financial or a technological return has been obtained and equals three if a financial as well as a technological return was achieved. A financial return is considered to be achieved if the ROI of the CVC investment is greater than or equal to the median ROI of all CVC. The mean (median) *return index* (Table 4.2) is 1.7 (2). We present the composition of return index in table 4.3 as a four-field matrix. We depict four possible groupings of CVC investments using technological and financial returns as two demarcation lines. Group I represents CVC investments resulting in low technological and low financial returns (271 dyads); Group II represents CVC investments resulting in low technological and high financial returns (285 dyads); Group III represents CVC investments resulting in high technological and low financial returns (70 dyads); and Group IV represents CVC investments resulting in high technological and high financial returns (60 dyads). Keep in mind that for dyads in Group I *return index* equals one; *return index* equals two if dyads are in either Group II or Group III; and, *return index* equals three if the dyad is in Group IV.

Table 4.3
Four Field Matrix of Technological and Financial Return
Note: Financial return equals 1 if ROI of CVC investment is equal or greater than the median ROI.

technological return (0) N=556	Group I N=271	Group II N=285
technological return (1) N=133	Group III N=73	Group IV N=60
	financial return (0) N=345	financial return (1) N=344

Table 4.3: Four Field Matrix of Technological and Financial Return

4.3.2.2 Independent Variables

We define several independent variables which we believe attempt to capture the timing of a CVC investment within a portfolio company's life cycle.

CVC round A-D/K. We define a binary variable as equal to one if the CVC invested in a particular financing round, zero otherwise. *CVC round A, CVC round B, CVC round C* and *CVC round D/K* represent investments made in round A, round B, round C or rounds D-K, respectively. As table 4.2 indicates, the investments are nearly equally split across rounds: Nearly 20% of the investments take place in round A, 21% in round B, 20% in round C and 16% in later rounds.

First CVC investment. In order to capture the first CVC investment in a particular biotechnology company we define the binary variable *first CVC investment* equal to one if the CVC investor in the focal CVC investment – dyad made the first CVC investment in a particular portfolio company, zero otherwise. Our investments are nearly split between first and subsequent investments; according to Table 4.2 55% of our CVC investments are first corporate (CVC) investments in a particular biotechnology company.

Company age at CVC investment. This variable depicts the age (in months) of the portfolio company at the time of the CVC investment. The mean (median) age at which a CVC invest in a portfolio company is 39 months (37 months).

4.3.2.3 Control Variables

We include several variables in our analysis, which are known or expected to influence the technological and financial return of a CVC. The variable *liquidation event=acquisition* controls for the mode of disinvestment and is equal to one if the first liquidation event of a portfolio company was an acquisition, zero otherwise. As financial returns are considered to be higher with an IPO, an acquisition is presumably negatively related to financial returns.[169] Ten percent of our investments result in an acquisition as first liquidity event, 86% in an IPO and less than 4% go bankrupt.

In line with the literature on VC investment[170], CVC experience can be considered as an indicator which proxies for the ability to select high quality portfolio companies and add value to the portfolio company. The variable *CVC stock* serves as proxy for CVC experience and depicts the number of investments a CVC investor has made prior to the focal investment. For our CVC firms, the mean (median) number of prior investments is 4 (2).

The variable *equity share* measures the equity share of a CVC investor at the time of investment. Since the value-adding capacity of CVC investors is limited, the value-adding activity should be positive correlated with their equity share. In addition, the larger their equity share, the more control rights the CVC may retain which is presumed to impact both types of CVC returns positively. According to Table 4.2 the average CVC takes a 17% stake in the biotechnology company.

The variable *valuation round A* is included as a proxy for the quality of the portfolio company and the level of information asymmetry present. Presumably, higher quality firms or technologies that are not as uncertain should be positively related with their initial round A valuations. The variable is measured as the valuation of a portfolio company at time of round A financing (in mil-

---

169 See Gompers (1995).
170 See for example Hsu (2004); Kaplan/ Schoar (2005); Chemmanur et al. (2008); Bottazi et al. (2007); Bengtson/ Sensoy (2008).

lion USD); the mean (median) round A financing amounts to 8 million USD (5 million USD).

The variables *alliance stock* and *patent stock* compute the total number of the portfolio companies' alliances and patents from founding up to the calendar month preceding the month in which the financial or technological returns are retained. There are conflicting views as to whether these stock variables should positively or negatively influence CVC returns. On the one hand patents and alliances reduce information asymmetries.[171] Therefore, investors might have to pay a premium when investing in companies with an already existing patent and alliance stock. Thus, the impact on *financial return* should be negative. On the other hand these stock variables reduce information asymmetries as they inform the CVC investors about the technological relatedness between the main business and the biotechnology (portfolio) company, and might therefore influence the *technological return* positively.[172] The average portfolio company entered 4 alliances prior to receiving CVC investment and received, on average, 3 patents from the USPTO.

Next, we include two variables that measure whether the biotechnology company received VC financing. The variable *VC investment* is equal to one if the biotechnology company received any VC funding, regardless of the time of that investment. The binary variable *VC Co-investment* measures whether any given round contains a joint investment between a VC and CVC. While the VCs primarily care about financial returns, CVCs, in addition to financial returns, also care about the strategic impact that the new venture will have on its core business.[173] Our presumption is that a

---

[171] See Haeussler et al. (2010), Hsu/ Ziedonis (2009) and Stuart et al. (1999).
[172] See Danzon et al. (2005).
[173] See Hellmann (2002).

joint investment with a VC increases the financial return but lowers the technological return of a CVC. We use *VC investment* as a control for the quality of the venture. For our sample 67% of the companies received VC financing and 17% of the funding rounds contained a joint investment between a VC and CVC.

Lastly, we include controls for the overall investment climate at the time of disinvestment. The variables *IPO activity* and *M&A activity* compute the number of U.S. IPOs and M&A transactions in the month of the first liquidation event. Similar to VC investments, CVC investments are subject to investment cycles, which implies that we should expect to see trends in annual investments and returns.[174] As returns are assumed to be greater in hot issue markets, *IPO activity* should have a positive impact on financial returns. In addition, the more M&A in a month, the more likely a venture might be acquired by its CVC investor.

The correlations are reported in Table 4.4. The relatively low correlation between *technological* and *financial return* suggest that these variables capture different dimensions of returns for CVC investors and indicate low complementarity between these two types of returns. In addition, few correlations show a correlation coefficient above 0.3. Not surprisingly *CVC round A* investment is correlated with *first CVC investment* (corr: 0.45) and the older the company *age at CVC investment*, the greater the *patent stock* in the month preceding the liquidation event (corr. 0.38).

---

174 See Gompers (2002).

Table 4.4
Correlations

	(1)	(2)	(3)	(4)	(5)	(6)	(7)	(8)	(9)
(1) Technological return	1								
(2) Financial return	0.07	1							
(3) Return index	0.60	0.15	1						
(4) CVC round A	-0.07	0.15	0.10	1					
(5) CVC round B	-0.09	-0.02	0.02	-0.25	1				
(6) CVC round C	0.01	-0.04	0.01	-0.25	-0.26	1			
(7) CVC round D-K	0.09	-0.03	0.00	-0.22	-0.23	-0.22	1		
(8) First CVC investment	0.04	0.08	0.12	0.45	0.11	-0.09	-0.13	1	
(9) Company age at CVC investment	0.07	0.01	-0.12	-0.37	-0.26	-0.01	0.15	-0.30	1
(10) Liquidation event=acquisition	-0.05	-0.04	-0.18	0.01	-0.01	-0.01	0.00	0.01	-0.08
(11) CVC stock	0.21	-0.01	0.09	-0.10	-0.09	-0.05	0.04	-0.14	0.18
(12) Equity stake	-0.11	0.17	0.05	0.39	0.12	-0.09	-0.17	0.27	-0.27
(13) Valuation round A	0.06	-0.10	0.02	0.09	0.03	0.04	-0.03	0.06	0.05
(14) Alliance stock	0.15	0.03	0.07	-0.23	-0.11	-0.01	0.10	-0.17	0.25
(15) Patent stock	0.09	-0.04	-0.04	-0.16	-0.15	-0.09	0.05	-0.14	0.38
(16) VC investment	0.07	-0.09	-0.04	-0.21	-0.03	0.02	0.02	0.06	0.06
(17) VC co-investment	-0.11	-0.03	-0.02	-0.03	0.13	0.00	-0.04	0.08	-0.10
(18) IPO activity	0.06	0.03	0.02	-0.01	-0.02	-0.03	0.08	0.05	0.05
(19) M&A activity	0.04	-0.02	0.02	-0.04	-0.03	-0.04	0.00	-0.02	0.15

	(10)	(11)	(12)	(13)	(14)	(15)	(16)	(17)	(18)
(11) CVC stock	-0.07	1							
(12) Equity stake	0.03	-0.17	1						
(13) Valuation round A	-0.17	0.06	-0.22	1					
(14) Alliance stock	-0.08	0.24	-0.21	0.14	1				
(15) Patent stock	-0.02	0.30	-0.18	0.04	0.17	1			
(16) VC investment	0.06	0.14	-0.11	-0.13	0.12	0.18	1		
(17) VC co-investment	0.18	0.02	-0.01	-0.02	-0.01	-0.05	0.32	1	
(18) IPO activity	-0.06	-0.07	0.04	0.04	0.01	-0.06	-0.04	-0.09	1
(19) M&A activity	0.09	0.13	-0.07	0.20	0.09	0.08	0.02	0.09	0.07

Table 4.4: Correlations

4.4 Empirical Results

4.4.1 Descriptive Statistics

Before we turn to our multivariate analysis, we relate our investment returns to the stage of CVC investment in Table 4.5, which suggests there appears to be a tendency towards later investments increasing technological returns. Whereas the percentage of realized technological return is 15% and 13% respectively, when the CVC investments occur in either round A or B, the percentage increases to 21% in round C and 26% in the rounds D-K. Presumably, uncertainty with regards to the technological fit between the portfolio company and the main business of the CVC is reduced in later round investments. While this seems to suggest that later stage investments result in more technology acquisitions, it also seems to be important to be the first CVC investor. We find that the first CVC investor manages to acquire the portfolio company's technology in 22% of investments, whereas follow-on CVC investors ultimately only acquire 18% of their portfolio technologies.[175]

Turning to financial return, we find a similar pattern in terms of the first and follow-on CVC investors but the opposite effect with regards to the financing rounds. The first CVC investor earns a median 53% annual ROI, whereas the follow-on CVC investors earn a median return of 36%.[176] Interestingly, ROI continuously drops from round A through round D-K. This suggests that the earlier the investment, the higher the financial return. While the univariate analysis offers already some insights in terms of a first entrant advantage, we now move on to test our presumptions in a multivariate setting allowing for *ceteri paribus* conclusions. We first run

---

175 Test of difference in means: p=0.16
176 Test of difference in means: p=0.004.

multivariate analysis with the return index, our joint measure of technological and financial return, and then separately for technological and financial return as dependent variables.

Table 4.5
Descriptive Statistics of CVC Timing and Returns
This table relates investment returns to the timing of the 689 CVC investments

Variable	N	Mean	P50	Std.Dev	Min	Max
Technological Return						
CVC round A	136	0.15	n.a.	n.a.	0	1
CVC round B	144	0.13	n.a.	n.a.	0	1
CVC round C	141	0.21	n.a.	n.a.	0	1
CVC round D-K	268	0.26	n.a.	n.a.	0	1
first CVC	376	0.22	n.a.	n.a.	0	1
follow on CVC	313	0.18	n.a.	n.a.	0	1
Total	689	0.20	n.a.	n.a.	0	1
Financial Return (ROI)						
CVC round A	136	6.21	0.81	31.55	-0.41	336.86
CVC round B	144	1.49	0.63	4.11	-0.35	45.14
CVC round C	141	0.75	0.46	1.26	-0.55	10.00
CVC round D-K	268	0.79	0.24	1.74	-0.73	11.99
first CVC	376	2.98	0.54	19.29	-0.72	336.86
follow on CVC	313	0.81	0.36	1.38	-0.73	9.00
Total	689	2.00	0.46	14.31	-0.73	336.86

Table 4.5: Descriptive Statistics of CVC Timing and Returns

4.4.2 Multivariate Regressions

4.4.2.1 Return Index

We present our regression using *return index* in table 4.6. Given that our dependent variable is a Likert score ordering (1, 2 and 3), we employ an ordered Logit model. We use clustered standard errors to account for any intra-firm correlations across the disturbances when one portfolio company appears more than one time in the dataset. In our sample, 20% of portfolio companies appear only one time, 29% appear two times and 51% appear more than two times.[177]

Baseline estimates are presented in model I of Table 4.6. Consistent with findings from prior studies, the composite return index is lower when the liquidation event is an acquisition opposed to an IPO[178], respectively bankruptcy. Additionally, we see a positive and significant relationship between the CVC investor's equity stake and our composite return index. However, this effect vanishes as soon as the independent variables are introduced. Composite returns are also larger the more alliances the portfolio company has entered. In correspondence with other studies, allying has positive effects on new product development[179] and investor returns[180]. Particularly, in biotechnology, the need for multiple and often complex capabilities as well as uncertainty surrounding the acceptance of new products have prompted companies to join alliances with

---

177 Note, that since the unit of observation is the CVC investor-company dyad, anytime more than one CVC investor invests in a portfolio company they will show up multiple times in the data.
178 See Gompers/ Lerner (2006).
179 See Rothaermel/ Deeds (2006).
180 See Haeussler (2006).

other firms to acquire the different capabilities[181] and signal quality[182].

Model II introduces the variables *CVC round A*, *CVC round B* and *CVC round C* with the reference category being *CVC round D/K*. In line with our expectations, the results suggest that the earlier the investment, the greater the pay-off for the CVC investor. In model III we include the *company age at CVC investment* to gain further insights into the potential time effect of investment. Again, we find that the earlier in the life cycle of a biotechnology company a CVC invests, the greater the impact on return index. Finally, in model IV, we include the dummy variable *first CVC investment* in order to measure whether the focal investment – dyad is related to the first CVC investor in a biotechnology company. This variable shows a highly significant and positive coefficient, which suggests that the first CVC investor earns significantly higher returns than a follow-on CVC investor.

Summing up, this first analysis suggests there is a premium for being the first CVC investor in a firm as well as that there is a first investor advantage in terms of the company life cycle in which an investor invests. In the following, we split our return index into technological and financial returns to get insights whether the determinants of technological and financial returns differ and whether the detected results in Table 4.5 are mainly driven by a specific return type or by both types.

---

[181] See Pisano (1990); Zahra (1996); Ahuja (2000).
[182] See Stuart et al. (1999).

Table 4.6
Return Index (Ordered Logit Model)

This table presents the multivariate regression results for the return index variable. An ordered logit model is employed, since the dependent variable is ordinal scaled. Clustered standard errors are used to control for portfolio company fixed-effects. Model I provides baseline estimates, model II to IV introduces the time variables. Standard errors are in parathesis. Asteriks indicate variables as being significant at 1%***, 5%**, and 10%* levels.

	Model I	Model II	Model III	Model IV
CVC round A		1.239***		
		(0.310)		
CVC round B		0.722***		
		(0.229)		
CVC round C		0.577**		
		(0.250)		
Company age at CVC investment			-0.0133***	
			(0.00316)	
First CVC investment				0.605***
				(0.164)
Liquidation event=acquisition	-1.265***	-1.295***	-1.366***	-1.277***
	(0.346)	(0.359)	(0.346)	(0.361)
CVC stock (ln)	0.218	0.249	0.247	0.268
	(0.164)	(0.184)	(0.173)	(0.172)
Equity share	1.031*	-0.0490	0.691	0.622
	(0.604)	(0.621)	(0.626)	(0.590)
Valuation round A (ln)	-0.0242	-0.131*	-0.0498	-0.0662
	(0.0721)	(0.0745)	(0.0745)	(0.0723)
Alliance stock (ln)	0.177*	0.367***	0.276**	0.240**
	(0.108)	(0.110)	(0.111)	(0.108)
Patent stock (ln)	-0.102	0.00620	0.0365	-0.0631
	(0.0957)	(0.102)	(0.105)	(0.0992)
VC investment	-0.210	-0.123	-0.246	-0.325*
	(0.173)	(0.174)	(0.175)	(0.176)
VC co-investment	0.168	0.0652	0.0819	0.156
	(0.237)	(0.251)	(0.230)	(0.247)
IPO activity	0.000624	0.00230	0.00191	0.000180
	(0.00402)	(0.00423)	(0.00409)	(0.00409)
M&A activity	0.0001	0.0002	0.0003	0.0001
	-0.0003	-0.0003	-0.0003	-0.0003
Observations	689	689	689	689
chi2	25.59	38.64	51.38	32.79
R-squared	0.0325	0.0498	0.0477	0.0431
ll	-615.4	-604.4	-605.8	-608.7
N_clust	355	355	355	355

Table 4.6: Return Index

4.4.2.2 Technological Return

We first isolate technological returns as dependent variable and present those results in table 4.7. We use a probit specification as our dependent variable *technological return* is a binary variable and use clustered standard errors to account for any intra-firms correlations across the disturbances when one biotechnology company appears more than one time in the dataset. Marginal effects are reported in the table. Again, model I (table 4.7) serves as our baselinfe model and contains only control variables. Comparing model I in table 4.6 and model I in table 4.7 we see some similarities and differences. For example, in both specifications alliance stock is positively and significantly related to returns. In contrast, CVC stock shows no significance in model I (table 4.6) suggesting that experienced CVCs, experience a greater technological return than less experienced CVCs. However, in this study we cannot distinguish whether this increase in returns is due to more experience in selecting the potential portfolio firms or whether they may be superior at nurturing these firms, or some combination of both effects. In terms of VC financing, we find that a VC financed biotechnology company (*VC investment*) has a positive and significant effect on *technological return* (table 4.7) but not *return index* (table 4.6). However, a co-investment in a given round between a VC and CVC (*VC co-investment*) negatively effects *technological return*. We will provide a more detailed interpretation of these results below when we present the financial return regressions (table 4.8).

Next, we include our variables focused on the timing of an investment. Interestingly, in model II (table 4.7), none of the individual round variables are significant. *Company age at CVC investment* in Model III is also not significant. Hence, the timing of the investment does not appear to affect a CVC investor's technological return. However, we find that being the first CVC investor is positively and highly significantly related to technological return (model IV). Presumably, the first CVC investor has more influence on the technological development and direction of the portfolio company than a follow-on investor. This first-mover advantage allows them to nurture a portfolio firm, build a more intimate relationship with them and also has an influence over the strategic direction of the firm.

In model V we re-run model III but in addition to controlling for biotechnology company fixed effects we also controlled for CVC investor fixed effects. We did this because 28% of our CVC investors have more than one dyadic relationship with our sample biotechnology companies. When comparing model III and model V we find that our results are robust to the inclusion of CVC investor fixed effects. We also re-ran our other models and found similar results. For brevity we include only model V.

Table 4.7
Technological Return (Probit)

This table presents the multivariate regression results for technological return. A probit model is employed, since the dependent variable is binary, providing marginal effects. Clustered standard errors are used to control for portfolio company fixed-effects. Model I provides baseline estimates, model II to IV introduces the time variables. In Model V also investor company fixed-effects are implemented. Standard errors are in parathesis. Asteriks indicate variables as being significant at 1%***, 5%**, and 10%* levels.

	Model I	Model II	Model III	Model IV	Model V
CVC round A		0.0711			0.258
		(0.0753)			(0.274)
CVC round B		-0.00172			-0.00670
		(0.0566)			(0.231)
CVC round C		0.0312			0.118
		(0.0459)			(0.181)
Company age at CVC investment			-0.000993		
			(0.000650)		
First CVC investment				0.119***	
				(0.0301)	
Liquidation event=acquisition	0.0227	0.0238	0.0205	0.0284	0.0896
	(0.0546)	(0.0548)	(0.0543)	(0.0553)	(0.201)
CVC stock (ln)	0.0714**	0.0712**	0.0728**	0.0785**	0.277**
	(0.0300)	(0.0304)	(0.0306)	(0.0315)	(0.113)
Equity stake	-0.0341	-0.0829	-0.0573	-0.121	-0.323
	(0.108)	(0.120)	(0.112)	(0.108)	(0.482)
Valuation round A (ln)	0.00955	0.00469	0.00733	0.00215	0.0183
	(0.0146)	(0.0149)	(0.0144)	(0.0141)	(0.0609)
Alliance stock (ln)	0.0845***	0.0910***	0.0911***	0.0961***	0.355***
	(0.0182)	(0.0194)	(0.0179)	(0.0181)	(0.0764)
Patent stock (ln)	0.00211	0.00592	0.0127	0.00758	0.0231
	(0.0160)	(0.0161)	(0.0155)	(0.0158)	(0.0643)
VC investment	0.0636*	0.0677**	0.0598*	0.0452	0.276*
	(0.0328)	(0.0330)	(0.0328)	(0.0336)	(0.151)
VC co-investment	-0.131***	-0.132***	-0.135***	-0.135***	-0.636***
	(0.0328)	(0.0341)	(0.0327)	(0.0324)	(0.214)
IPO activity	0.000858	0.000922	0.000949	0.000756	0.00359
	(0.000706)	(0.000714)	(0.000710)	(0.000687)	(0.00279)
M&A activity	-1.59e-05	-1.03e-05	-7.98e-06	-1.73e-05	-4.01e-05
	(6.28e-05)	(6.34e-05)	(6.22e-05)	(6.30e-05)	(0.000273)
Constant					-1.965***
					(0.325)
Observations	689	689	689	689	689
r2_p	0.112	0.115	0.115	0.132	
ll	-307.8	-306.7	-306.5	-300.7	-306.7
chi2	69.68	69.52	74.93	81.17	
N_clust	355	355	355	355	

Table 4.7: Technological Return

4.4.2.3 Financial Return

Finally, we present our results using *financial return* as the dependent variable in table 4.8. In this specification we use an ordinal linear regression given our variable is the logarithmic form of the ROI and, analogous to the former specifications, we include biotechnology company fixed effects.

In our baseline model I (table 4.8) with only control variables, we find, in line with our expectations, that *financial return* is significantly lower, when the liquidation event is an acquisition. When the liquidation event is an acquisition, the ROI is reduced by 43.6% compared to the ROI when the liquidation event is an IPO. In the specifications presented in model table 4.8 we also include as a control the binary variable *license* which equals one, if the CVC investment came along or was followed by a licensing agreement, zero otherwise. We presume that a license agreement between the CVC investor and the portfolio company might have negative effects on financial returns. The CVC investor might use its influence in order to negotiate for a more favourable licensing deal which may, potentially, increase the costs of the biotechnology company thereby reducing ROI. Indeed, we find that a technology acquisition by the CVC investor in the form of a license agreement results in an 11% reduction in ROI compared to a dyadic relationship without a licensing agreement.

Whereas *patent stock* shows no impact on *technological return*, we find a weak negative relationship with *financial return*. Presumably, patents increase the bargaining power of biotechnology companies and thus result in higher company valuation at the time of CVC investment thereby reducing the financial return to the

CVC investors.[183] Interestingly, we find a negative impact on *financial return* for the CVC when the biotechnology received VC financing. Two explanations might apply. First, it might be that the CVC invested under less favourable conditions compared to when a VC invested in the company. The VC might have some ability to increase the valuation of the company thereby having the CVC pay either more or getting a lower share in the company. Second, a later-stage incoming VC might negotiate hard and dilute the share of the CVC. While it is common practice in VC agreements to include anti-dilution provisions, if they are not as common in CVC agreements then we could see this effect.

When we turn to the variable *VC co-investment* we find that CVC investors who co-invest with a VC, in the same round, receive a ROI which is 15% higher than when the CVC invests alone. This result might be once more due to the negotiation power of VCs and presumably, in addition, to the primary financial focus of the VC firm. In line with our expectations, we find the opposite when we compare this result to our regressions in table 4.7: VC syndication results in a lower technological return but into a higher financial return for the CVC investor.

---

183 See Hsu/ Ziedonis (2008).

Table 4.8
Financial Return (Ordinally Least Squares)

This table presents the multivariate regression results for financial return. An ordinal linear regression model is employed, since the dependent variable is the logaritmic form of the ROI. Clustered standard errors are used to control for portfolio company fixed-effects. Model I provides baseline estimates, model II to IV introduces the time variables. In Model V also investor company fixed-effects are implemented. Standard errors are in parathesis. Asteriks indicate variables as being significant at 1%***, 5%**, and 10%* levels.

	Model I	Model II	Model III	Model IV	Model V
CVC round A		0.435***			0.435***
		(0.124)			(0.125)
CVC round B		0.212**			0.212**
		(0.0904)			(0.0902)
CVC round C		0.0627			0.0627
		(0.0663)			(0.0632)
First CVC investment			-0.00345**		
			(0.00139)		
Company age at CVC investment				0.131**	
				(0.0640)	
Liquidation event acquisition	-0.436***	-0.430***	-0.452***	-0.433***	-0.430***
	(0.0764)	(0.0785)	(0.0754)	(0.0771)	(0.0755)
CVC stock (ln)	-0.0264	-0.0215	-0.0202	-0.0153	-0.0215
	(0.0355)	(0.0352)	(0.0348)	(0.0358)	(0.0367)
License	-0.106*	-0.125**	-0.120*	-0.130**	-0.125*
	(0.0605)	(0.0565)	(0.0616)	(0.0603)	(0.0672)
Equity stake	0.916**	0.522	0.832**	0.820**	0.522*
	(0.365)	(0.329)	(0.375)	(0.353)	(0.304)
Valuation round A (ln)	-0.0450	-0.0791**	-0.0512*	-0.0535*	-0.0791**
	(0.0291)	(0.0346)	(0.0297)	(0.0305)	(0.0341)
Alliance stock (ln)	-0.00947	0.0505	0.0151	0.00455	0.0505
	(0.0473)	(0.0562)	(0.0480)	(0.0506)	(0.0472)
Patent stock (ln)	-0.0555**	-0.0256	-0.0209	-0.0482*	-0.0256
	(0.0262)	(0.0290)	(0.0298)	(0.0267)	(0.0294)
VC investment	-0.193**	-0.156*	-0.201**	-0.213***	-0.156**
	(0.0823)	(0.0795)	(0.0795)	(0.0819)	(0.0761)
VC co-investment	0.152**	0.119*	0.128**	0.142**	0.119*
	(0.0604)	(0.0626)	(0.0610)	(0.0602)	(0.0681)
IPO activity	-0.000161	0.000312	0.000164	-0.000223	0.000312
	(0.00111)	(0.00117)	(0.00111)	(0.00111)	(0.00119)
M&A activity	0.000232*	0.000257*	0.000270**	0.000232*	0.000257**
	(0.000133)	(0.000131)	(0.000130)	(0.000133)	(0.000130)
Constant	(0.119)	(0.136)	(0.125)	(0.125)	(0.139)
	0.499***	0.318**	0.564***	0.442***	0.318**
Observations	689	689	689	689	689
R-squared	0.131	0.160	0.144	0.137	0.160
ll	-699.4	-687.5	-694.1	-696.7	-687.5
F	5.890	6.003	7.889	5.770	6.371
N_clust	355	355	355	355	577

Table 4.8: Financial Return

In model II, we add our CVC round variables. The results suggest that investing early pays off. When a CVC invests in round A, its ROI is 44% higher than an investment in round D or later. The size of the effect is reduced by about 50% when the CVC invests in round B and the ROI is still 21% higher than when the CVC invests in round D or later. This time effect diminishes when we compare the pay-off in round C to later rounds (model III). Our second measure for time effects, *company age at CVC investment*, mirrors this timing effect. The earlier the CVC invests, the higher its financial return. Lastly, we turn to our variable indicating whether the CVC was the first investing CVC or a follower (*first CVC investor*). In line with our expectations, we find that the first CVC again receives a discovery premium and experiences a first-mover advantage. In terms of size effects, the first CVC receives a 13% higher return than follow-on CVC investors. Analogous, in model IV, we re-ran as a robustness check model II taking into account CVC investment company fixed effects; similar to model V (table 4.7), the effects remain robust.

4.5 Conclusion

The timing of investment is an important determinant when analysing returns occurring to CVC investors. According to corporate finance theory, risk-averse investors should receive a premium for investing in earlier stages of a company's life cycle due to higher risk in this stage. In contrast, empirical studies on VC investment report higher returns for investments in later stages of a portfolio company's life cycle. The misalignment between theory and empirics suggest that either theory is wrong or the VC setting is a very special field of analysis in which either the first-mover advantage is absent or compensated through other forces, or some other unidentified distortion exists.

In this study, we test whether the first-mover argument applies for CVC investors. In contrast to VC investors who seek only financial returns, CVC investors also have strategic incentives to invest in nascent, research-intensive companies.

We argue that three sources of early- and first-mover advantages are apparent: First, in line with corporate finance theory, first-movers receive a premium as compensation for accepting greater risk. Second, first-movers receive a discovery premium due to their superior capability in identifying a promising venture. Third, being a first-mover enables the CVC investor to set the portfolio company on a technology path that is aligned with their own business units and thus, CVC investing is an attractive mechanism for technology sourcing.

Overall, our multivariate analysis provides strong support for our notion of a first-mover advantage in gaining both technological and financial returns.

Moreover, we also identified a more general early investor advantage which exists for financial but not technological returns. This seems to indicate that the first-mover CVC is unable to capture full rents through their investment and that fast-followers or other early investors are able to capture some of the financial return generated by the portfolio firm. If for example, a CVC investor was solely interested in financial returns, similar to a VC investor, they could follow a simple investing rule: identify CVC investments that are being made for technological reasons and make a fast, follow-on CVC investment. As these portfolio firms progress, the value of their firm will increase thereby increasing the return to other early CVC investors. Interestingly, we do not see this same kind of effect when we focus on technological returns. In that case, first-movers clearly have an advantage. Although we did suggest that it might be possible that follow-on CVC investors experienced some indirect R&D spill overs, which we do not identify.

The study is subject to a few limitations. First, there is a question of external validity. This study investigated the presence of a first-mover advantage for CVC investors in the bio-pharmaceutical industry. The bio-pharmaceutical industry is characterized by a highly uncertain and complex R&D process with companies having to access a broad range of human resources and capital to be successful in product development. Young biotechnology companies often have promising ideas but lack the financial resources and complementary assets to turn a promising idea into a marketable product. On the other hand, incumbent firms are challenged by patent expiration of their blockbuster drugs, increasing R&D expenses and shrinking product pipelines. External sourcing activities such as CVC are increasingly important for these firms to sustain a competitive advantage. While we assume that our general mechanisms in terms of a first-mover advantage are also present in industries other than biotechnology, future work is needed to validate our results in other contexts. Second, we conjecture that the source of a first-mover advantage is mainly due to a discovery premium due to superior selection abilities and the overall ability of the first-mover to particularly influence the strategic path of the portfolio company. While we assume that both notions apply, our analysis does not allow us to distinguish the size of the effects on the technological or financial return. This could be a promising avenue of future research. In conclusion, we believe that our study opens up many exciting vistas for future work, and we seek to stimulate further theoretical refinement and empirical investigation in furthering our understanding of this apparent first-mover advantage for CVC investors.

Bibliography for Chapter 4

Ahuja, Manju (2000): Collaboration Networks, Structural Holes and Innovation: A Longitudinal Study, in: Administrative Science Quarterly 45, 425-453.

Allen, Stephen / Hevert, Kathleen (2007): Venture Capital Investing by Information Technology Companies: Did It Pay? in: Journal of Business Venturing 22, 262-282.

Alpert, Frank / Kamins, Michael / Graham, John (1992): An Examination of Reseller Buyer Attitudes Toward Order of Brand Entry, in: Journal of Marketing 56, 25-37.

Asset Alternatives (2002): The Corporate Venturing Directory and Yearbook, Asset Alternatives, New York.

Baldwin, William / Childs, Gerald (1969): The Fast Second and Rivalry in Research and Development, in: Southern Economic Journal 36, 18-24.

Baum, Joel / Silverman, Brian (2004): Picking Winners or Building Them? Alliance, Intellectual, and Human Capital as Selection Criteria in Venture Financing and Performance of Biotechnology Start-ups, in: Journal of Business Venturing 19, 411-436.

Bengtson, Ola / Sensoy, Berk A. (2008): Investor Abilities and Financial Contracting: Evidence from Venture Capital, Working Paper, Cornell University, August 2008.

Benson, David / Ziedonis, Rosemarie (2010): Corporate Venture Capital and the Return to Acquiring Portfolio Companies, in: Journal of Financial Economics 98, 478-499.

Bottazi, Laura / Da Rin, Marco / Hellmann, Thomas (2008): Who Are the Active Investors? Evidence from Venture Capital, in: Journal of Financial Economics 89, 488-512.

Brealey, Richard / Myers, Steward / Allen, Franklin (2006): Corporate Finance, 8th Edition, McGraw-Hill.

Burton, Diane / Sorensen, Jesper / Beckman, Christine (2002): Coming from Good Stock: Career Histories and New Venture Formation, in: Research in the Sociology of Organizations 19, 229-262.

Carpenter, Gregory / Nakamoto, Kent (1989): Consumer preference formation and pioneering advantage, in: Journal of Marketing Research 26, 285-298.

Chandler, Alfred (1977): The Visible Hand: The Managerial Revolution in American Business, Cambridge, MA, Harvard University Press.

Chemmanur, Thomas / Fulghieri, Paolo (1994): Investment Bank Reputation, Information Production, and Financial Intermediation, in: Journal of Finance 48, 57-79.

Chemmanur, Thomas / Krishnan, Karthik / Debarshi, Nandy (2008): How Does Venture Capital Financing Improve Efficiency in Private Firms? A Look Beneath the Surface, Working Paper, Carroll School of Management, February 2008.

Chesbrough, Henry W. (2002): Making Sense of Corporate Venture Capital, in: Harvard Business Review March 2002, 90-99.

Corporate Strategy Board (2000): Corporate Venture Capital: Managing Equity Investments for Strategic Returns, Unpublished Working Paper, Corporate Strategy Board.

Cumming, Douglas (2008): Contracts and Exits in Venture Capital Finance, in: The Review of Financial Studies 21, 1947-1982.

Cumming, Douglas / Walz, Uwe (2004): Private Equity Returns and Disclosure Around the World, LSE Working Paper, April 2004.

Das, Sanjiv / Jagannathan, Murali / Atulya, Sarin (2003): Private Equity Returns: An Empirical Examination of the Exit of Venture-Backed Companies, in: Journal of Investment Management 1, 1-26.

Danzon, Patricia M. / Nicholson, Jean / McCullough, Jeffrey (2005): Biotech-Pharmaceutical Alliances as a Signal of Asset and Firm Quality, in: Journal of Business 78, 1433-1464.

Dushnitsky, Gary (2006): Corporate Venture Capital: Past Evidence and Future Directions, in: Mark Casson et al. (ed.), The Oxford Handbook of Entrepreneurship, Oxford University Press, 387-431

Dushnitsky, Gary / Lenox, Michael (2005): When Do Firms Undertake R&D by Investing in New Ventures?, in: Strategic Management Journal 26, 947-965.

Dushnitsky, Gary / Lenox, Michael (2006): When Does Corporate Venture Capital Investment Create Firm Value?, in: Journal of Business Venturing 21, 723-772.

Dushnitsky, Gary / Shapira, Zur (2009): Entrepreneurial Finance Meets Organizational Reality: Comparing Investment Practices by Corporate and Independent Venture Capitalists, Working Paper, University of Pennsylvania, August 2009.

Hand, John (2007): Determinants of the Round-to-Round Returns to Pre-IPO Venture Capital Investments in U.S. Biotechnology Companies, in: Journal of Business Venturing 22, 1-28.

Eisenhardt, Kathleen / Schoonhoven, Claudia-Bird (1990): Organizational Growth: Linking Founding Team, Strategy, Environment, and Growth Among U.S. Semiconductor Ventures, 1978-1988, in: Administrative Science Quarterly 35, 504-529.

Gal-Or, Esther (1985): First Mover and Second Mover Advantages, in: International Economic Review 26, 649-653.

Gilbert, Richard / Newberry, David (1982): Preemptive Patenting and the Persistence of Monopoly, in: American Economic Review 72, 514-525.

Golder, Peter / Tellis, Gerard (1993): Pioneer Advantage: Marketing Logic or Marketing Legend?, in: Journal of Marketing Research 30, 158-170.

Gompers, Paul (1995): Optimal Investment, Monitoring, and Staging of Venture Capital, in: Journal of Finance 50, 1461-1489.

Gompers, Paul / Lerner, Josh (2000): The Determinants of Corporate Venture Capital Success: Organizational Structure, Incentives, and Complementaries, in: Randall Morck (ed.), Concentrated Ownership, University of Chicago Press, 17-54.

Gompers, Paul (2002): Corporations and the Financing of Innovation: the Corporate Venture Experience, in: Federal Reserve Bank of Atlanta Economic Review 87, 1-18.

Gompers, Paul / Lerner, Josh (2006): The Venture Capital Cycle, 2nd Edition, MIT Press.

Haeussler, Carolin (2006): When does Partnering create Market Value? A Transaction Cost and Signaling Theory Approach, in: European Management Journal 24, 1-15.

Haeussler, Carolin / Harhoff, Dietmar / Müller, Elisabeth (2009): To Be Financed or Not - The Role of Patents for Venture Capital Financing, Discussion Paper 2009-02 Munich School of Management, January 2009.

Haines, Daniel / Chandran, Rajan / Parkhe, Arvinde (1989): Winning By the First to Market...or Second?, in: Journal of Consumer Marketing 6, 63-69.

Hannan, Michaeal / Burton, Diane / Baron, James (1996): Inertia and Change in the Early Years: Employment Relations in Young, High Technology Firms, in: Industrial and Corporate Change 5, 503-536.

HBM Pharma (2009): Biotech M&A Survey, Trade Sales of Biotechnology and Specialty Pharma Companies 2005-2008, February 2009.

Hege, Ulrich / Palomino, Frédéric / Schwienbacher, Armin (2003): Determinants of Venture Capital Performance: Europe and the United States, Working Paper University of Amsterdam, November 2003.

Hellmann, Thomas (2002): A Theory of Strategic Venture Investing, in: Journal of Financial Economic 64, 285-314.

Hsu, David (2004): How Much Do Entrepreneurs Pay for Venture Capital Affiliation, in: Journal of Finance 59, 1805-1844.

Hsu, David / Ziedonis, Rosemarie (2008): Patents as Quality Signals for Entrepreneurial Ventures, in: Academy of Management Best Paper Proceedings.

Ivanov, Vladimir / Xie, Vei (2010): Do Corporate Venture Capitalists Add Value to Start-up Firms? Evidence from IPOs and Acquisitions of VC-Backed Companies, in: Financial Management 39, 129-152.

Kang, Daniel H. / Nanda, Vikram (2010): Complements or Substitutes? Technological and Financial Returns Created from Corporate Venture Capital Investments, Working Paper, Georgia Institute of Technology, June 2010.

Kaplan, Steven / Schoar, Antoinette (2005): Private Equity Performance: Returns, Persistence, and Capital Flows, in: Journal of Finance 60, 1791-1823.

Kardes, Frank / Kalyanaram, Gurumurthy (1992): Order-of-Entry Effects on Consumer Memory and Judgement: An Information Integration Perspective, in: Journal of Marketing Research 29, 343-357.

Kardes, Frank / Kalyanaram, Gurumurthy / Chandrashekaran, Murali / Dornoff, Ronald (1993): Brand Retrieval, Consideration Set Composition, Consumer Choice, and the Pioneering Advantage, in Journal of Consumer Research 20, 62-75.

Lavie, Dovev / Lechner, Christoph / Singh, Harbir (2007): The Performance Implications of Entry and Involvement in Multipartner Alliances, in Academy of Management Journal 50, 589-604.

Li, Yong / James, Barclay / Madhavan, Ravi / Mahoney, Joseph (2006): Real Options: Taking Stock and Looking Ahead, in: Advances in Strategic Management 24, 31-66.

Lieberman, Marvin / Montgomery, David (1988): First-Mover Advantages, in: Strategic Management Journal 9, 41-58.

Lieberman, Marvin / Montgomery, David (1998): First-Mover (Dis)advantages: Retrospective and Link With the Resource-Based View, in: Strategic Management Journal 19, 1111-1125.

Manigart, Sophie / De Waele, Koen / Wright, Mike / Robbie, Ken / Desbrie`res, Philippec / Sapienza, Harry J./ Beekman, Amy (2002): Determinants of Required Return in Venture Capital Investments: A Five Country Study, in: Journal of Business Venturing 7, 291-312.

Maula, Markku / Autio, Erkko / Murray, Gordon (2005): Corporate Venture Capitalists and Independent Venture Capitalists: What do They Know, Who do They Know and Should Entrepreneurs Care?, in: Venture Capital 7, 3-21.

Maula, Markku / Murray, Gordon (2002): Corporate Venture Capital and the Creation of U.S. Public Companies: the Impact of Sources of Venture Capital on the Performance of Portfolio Companies, in: Hitt / Amit / Lucier / Nixon (Eds.), Creating Value: Winners in the New Business Environment, Blackwell Publishers, Oxford, 164-187.

MacMillan, Ian / Roberts, Edward / Livada, Val / Wang, Andrew (2008): Corporate Venture Capital (CVC): Seeking Innovation and Strategic Growth, National Institute of Standards and Technology, U.S. Department of Commerce, June 2008.

Pisano, Gary (1990): The R&D Boundaries of the Firm: An Empirical Analysis, in: Administrative Science Quarterly, Vol. 35, pp. 153-176.

PriceWaterhouseCoopers (2006): Corporate Venture Capital Activity on the Rise in 2006, PriceWaterhouseCoopers, Washington DC.

Rothaermel, Frank / Deeds David (2006): Alliance Type, Alliance Experience and Alliance Management Capability in High-Technology Ventures, in: Journal of Business Venturing 21, 429-460.

Schildt, Henry / Maula, Markku / Keil, Thomas (2005): Explorative and exploitative learning from external corporate ventures, in: Entrepreneurship Theory and Practice 29, 493-515.

Schmalansee, Richard (1982): Product Differentiation Advantages to Pioneering Brands, in: American Economic Review 72, 159-180.

Shane, Scott / Stuart, Toby (2002): Organizational Endowments and the Performance of University Start-Ups, in: Management Science 48, 154-170.

Smith, Sheryl (2009): The Company They Keep: Innovation Returns to Corporate Venture Capital in the Medical Device Industry, Entrepreneurial Clinicians, and Competitive Convestors, Working Paper, Temple University, July 2009.

Söderblom, Anna / Wiklund, Johan (2006): Factors Determining the Performance of Early Stage High-Technology Venture Capital Funds – A Review of the Academic Literature, Small Business Service, March 2006.

Spence, Michael (1984): Cost Reduction, Competition, and Industry Performance, in: Econometrica 52, 101-122.

Stuart, Toby / Hoang, Ha / Hybels, Ralph (1999): Interorganizational Endorsements and the Performance of Entrepreneurial Ventures, in: Administrative Science Quarterly 44, 315-349.

Teece, David (1980): Economics of Scope and the Scope of an Enterprise, in: Journal of Economic Behavior and Organization 10, 223-247.

Wadhwa, Anu / Kotha, Suresh (2006): Knowledge Creation through External Venturing: Evidence From the Telecommunications Equipment Manufacturing Industry, in: Academy of Management Journal 49, 819-835.

Wadhwa, Anu / Phelps, Corey (2010): An Option to Ally: A Dyadic Analysis of Corporate Venture Capital Relationships, SSRN Working Paper, February 2010.

Wernerfelt, Birger / Karnani, Aneel (1987): Competitive Strategy Under Uncertainty, in: Strategic Management Journal 8, 187-194.

Zahra, Shaker (1996): Governance, Ownership, and Corporate Entrepreneurship: The Moderating Impact of Industry Technological Opportunities, in: Academy of Management Journal 39, 1713-1735.

Armin Höll-Steier

Venture Capital
Fund Certification, Performance Prediction and Learnings from the Past

Frankfurt am Main, Berlin, Bern, Bruxelles, New York, Oxford, Wien, 2011.
XXVI, 243 pp., num. tab. and graphs.
European University Studies. Series 5: Economics and Management. Vol. 3381
ISBN 978-3-631-61834-9 · pb. € 52,80*

This book contains three studies. The first study investigates the relationship between private equity investors and fund managers and how intermediaries can mitigate their agency problems. The incentive structure of three intermediary types and their behavior in signaling fund qualities to investors are studied theoretically. A recommendation which intermediary to consult is given. The second study presents a new statistical method to predict the performance distribution of venture capital direct investments. The accuracy of this method is investigated and compared to existing approaches. The third study is about the European venture capital market's historic development before and after the internet bubble and reasons for the bad development especially after the bubble.

Content: External certification in the fundraising of first-time private equity funds · A new approach to predict the performance of venture capital direct investments · European venture capital: What can we learn from the past?

Frankfurt am Main · Berlin · Bern · Bruxelles · New York · Oxford · Wien
Auslieferung: Verlag Peter Lang AG
Moosstr. 1, CH-2542 Pieterlen
Telefax 0041 (0) 32/376 17 27

*inklusive der in Deutschland gültigen Mehrwertsteuer
Preisänderungen vorbehalten
Homepage http://www.peterlang.de